P9-CMZ-600

"[*The Serengeti Rules*] is wholeheartedly recommended for its entertaining view of biology from an original perspective."

—*Financial Times*

"A compelling read filled with big, bold ideas."

—Brian J. Enquist, *Nature*

"The rules Carroll establishes can guide us as we come to the realization that conservation is a series of choices we make every day, deciding what to save and what will slip away."

—*Wall Street Journal*

"A thought-provoking challenge to complacency."

—*Kirkus*

"Carroll's book is fantastic, a success story in going form the specific to the general. It helps that Carroll is a gifted writer, captivating and thoughtful, and highly respectful of the reader."

—Greg Laden, *ScienceBlog*'s Greg Laden

"Sean B. Carroll's new book *The Serengeti Rules* is a passionate telling of the story of the precarious and hard-fought balance that is the very precondition of health—both at the level of individual organisms and at the level of ecosystems....The book is informative, well-written, and persuasive....*The Serengeti Rules* is an optimistic book."

—Alva Noë, NPR.org's 13.7 blog

"[A] triumphant account of how physiology and ecology turned out to share some of the same mathematics."

—Simon Ings, *New Scientist*

"[A] deep journey into the rules of life on Earth."

—Cathy Taibbi, Examiner.com

"*The Serengeti Rules* should be widely read."

—Neil Paterson, *Dundee University Review of the Arts*

"Sean Carroll's new book, with his thesis that everything is regulated backed by stories of discovery and inquiry, will enhance the way I teach biology. I am convinced that *The Serengeti Rules* should be required reading for students in all fields of science, but especially those pursuing careers in biology education."

—Paul K. Strode, *American Biology Teacher*

"This book offers hope that we can make a difference, that we can follow those rules, and that things can get better on our planet, our home. It is well written, meticulously researched, and easy to read. I also learned more about the serendipitous nature of scientific discovery. I thoroughly enjoyed this book and highly recommend it to both teachers and students."

—Cheryl Hollinger, *American Biology Teacher*

"This book was easy to read and gave many great examples of the resiliency of nature."

—National Science Teachers Association Recommends

"[*The Serengeti Rules*] successfully conveys a powerful message: although biology is infinitely complex and diverse, simple sets of rules of regulation that apply across scales, from molecules to the entire planet's ecosystem, can and have been identified. They are also remarkably easy to explain, as shown by the many beautiful examples described in the book....[A] great read."

—*Cell*

"In this remarkably engaging book, Carroll...persuasively argues that life at all levels of complexity is self-regulated, from the inner workings of cells to the larger relationships governing the Serengeti ecosystem....Carroll superbly animates biological principles while providing important insights."

—*Publishers Weekly*

"*The Serengeti Rules* is one of the best biology books for general readers I've ever encountered. It should be required reading for every college student, regardless of major."

—Andrew H. Knoll, Harvard University

"A master storyteller, Carroll explores the unity of biology from the molecular level to the Serengeti, the rules that regulate life, and the consequences when regulation breaks down. A fascinating journey from beginning to end, this book will educate and entertain readers at all levels and leave them with a better understanding of how the biosphere works."

—Simon Levin, Princeton University, author of
Fragile Dominion: Complexity and the Commons

"This is a rattling good read by one of the leading scientists of our time. *The Serengeti Rules* made me think differently about what we biologists do. This is a book that needs to be shouted from the rooftops."
—Andrew F. Read, Pennsylvania State University

"Masterful and compelling. *The Serengeti Rules* is a significant contribution, one that will be welcomed by professional biologists and a wide range of lay readers."
—Harry W. Greene, author of *Tracks and Shadows: Field Biology as Art*

THE SERENGETI RULES

THE SERENGETI RULES

THE QUEST TO DISCOVER HOW LIFE WORKS AND WHY IT MATTERS

SEAN B. CARROLL

PRINCETON UNIVERSITY PRESS
Princeton & Oxford

Requests for permission to reproduce material from this work
should be sent to Permissions, Princeton University Press

Published by Princeton University Press
41 William Street, Princeton, New Jersey 08540

In the United Kingdom: Princeton University Press
6 Oxford Street, Woodstock, Oxfordshire OX20 1TR

press.princeton.edu

Cover design by Chris Ferrante
Cover photograph © TimFitzharris.com

Fourth printing, first paperback printing, 2017

Cloth ISBN 978-0-691-16742-8

Paper ISBN: 978-0-691-17568-3

Library of Congress Control Number: 2015038116

British Library Cataloging-in-Publication Data is available

This book has been composed in Sabon Next LT Pro and League Gothic

Printed on acid-free paper. ∞

Printed in the United States of America

7 9 10 8

For the animals,
and the people looking out for them

Suppose it were perfectly certain that the life and fortune of every one of us would, one day or other, depend upon his winning or losing a game of chess. Don't you think that we should all consider it to be a primary duty to learn at least the names and the moves of the pieces? . . . Yet it is a very plain and elementary truth that the life, the fortune, and the happiness of every one of us, and, more or less, of those who are connected with us, do depend upon our knowing something of the rules of a game infinitely more difficult and complicated than chess. It is a game which has been played for untold ages. . . . The chessboard is the world, the pieces are the phenomena of the universe, the rules of the game are what we call the laws of Nature.

—THOMAS H. HUXLEY, *A LIBERAL EDUCATION* (1868)

CONTENTS

THE SERENGETI RULES

FIGURE 1 The Naabi Entrance Gate, Serengeti National Park.

Photo courtesy of Patrick Carroll.

MIRACLES AND WONDER

The corrugated gravel road known officially as Tanzania Route B144 provides a bone-jarring, teeth-rattling, bladder-testing connection between two of the great wonders of Africa.

At its eastern end stand the massive green slopes of Ngorogoro Crater, a giant, more than ten-mile-wide caldera formed by the collapse of one of the many extinct volcanoes of the Great Rift Valley, and home to more than 25,000 large mammals. To the west lie the vast plains of the Serengeti, our destination on this cloudless, postcard-perfect day.

The route in between is a stark contrast to the lush Ngorogoro highlands. There is no visible source of water; the Maasai herdsmen and boys we pass in their bright red shuka graze their livestock on whatever brown stubble they can find. But as we bounce our way through the first simply marked gate to Serengeti National Park, the landscape changes.

The Maasai vanish, and the nearly barren tracts they use are replaced by straw-colored grasslands, and instead of cattle and goats, sleek black-striped Thomson gazelles look up to see who or what is kicking up dust all over their breakfast.

The anticipation in our Land Cruiser rises. Where there are gazelles, there may be other creatures lurking in the tall grass. We pop open the top of the vehicle, stand up, and with the African rhythms of Paul Simon's *Graceland* playing in my head, I start to scan back and forth. This is my first visit to what the Maasai call "Serengit" for "endless plains." Joining me on my pilgrimage to this legendary wildlife sanctuary is my family:

> *pilgrims with families and we are going to Graceland . . .*

At first, I am a bit concerned. Where is all the wildlife? Yes, it is the dry season, but things look *really dry.* Can this place live up to its reputation?

The continuous grass plain is broken only occasionally by small rocky hills, or *kopjes.* From their granite boulders, animals (or tourists) can scan around for miles. There are also gray or red termite mounds projecting up to a few feet over the tops of the grass. One's eye is naturally drawn to these shapes.

"What is that over there?" asks a voice in the vehicle.

A couple of us grab our binoculars and zero in on a lone mound a couple of hundred yards away.

"Lion!"

A golden lioness is standing on top, staring out over the surrounding grass.

OK, so they are here, I murmur to myself. *But this is the famous Serengeti?*

It is going to be really hard to spot things in this tall dry grass. I am the only biologist in my clan, I can't expect anyone else to want to do this for days on end.

As we drive on, some streaks of green grass appear, with a few iconic flat-topped acacia trees sprinkled about. A creek bed meanders through the green patches, and it has plenty of water. We go over a small rise, round a bend, and skid to a stop—zebra and wildebeest block the road and fill the entire view.

It is a sea of stripes. Perhaps 2,000 or more animals have gathered near a large waterhole, raising a ruckus. The zebras' calls are something between a bark and a laugh: "kwa-ha, kwa-ha," while the wildebeest seem to just mutter "huh?" These herds are stragglers from the greatest animal migration on the planet, when as many 1 million wildebeest, 200,000 zebras, and tens of thousands of other animals follow the rains north to greener grazing grounds.

Coming next to the waterhole from over the small rise on our left—the Dawn Patrol—a parade of elephants with several youngsters scurrying to keep up. The herds part to make way.

From that point on, the Serengeti offers an unending canvas containing mammals of many sizes, shapes, and colors: small gray warthogs with tails standing straight up like our radio antenna; not two or three but at least nine species of antelope—the tiny dik-dik, the massive eland, impala, topi, waterbuck, hartebeest, Thomson's and the larger Grant's gazelles, and the ubiquitous wildebeest; black-backed jackals; towering Masai giraffe; and yes, all three big cats on this first day, including several more lions, a leopard dozing in a tree, and a cheetah posing just feet from the road.

Although I have seen many pictures and movies, nothing prepared me for, nor spoiled the thrill of, encountering this stunning scenery for the first time.

A strange, but very pleasant feeling sweeps over me as I gaze across a wide green valley, with multitudes of creatures and acacia stretching as far as I can see, and the sun beginning to set behind the silhouettes of the surrounding foothills. Although it is the first time I have ever been to Tanzania, I feel at *home*.

And indeed, this is home. For across the Rift Valley of East Africa lay buried the bones of my and your ancestors, and those of our ancestors' ancestors. Sandwiched between Ngorogoro Crater and the Serengeti lies Olduvai Gorge, a thirty-mile-long twisting maze of badlands. It was in its eroding hillsides (just three miles off of the current B144) that, after decades of searching, Mary and Louis Leakey (and their sons) unearthed not one, not two, but *three* different species of hominids that had lived in East Africa 1.5 to 1.8 million years ago. Thirty miles to the south at Laetoli, Mary and her team later discovered 3.6-million-year-old footprints made by our small-brained but upright-walking ancestor *Australopithecus afarensis*.

Those hard-earned hominid bones were precious needles in a haystack of other animal fossils that tell us that, although the specific actors have changed, the drama we can still see today—of fleet herds of grazing animals trying to stay out of the reach of a number of wily predators—has been playing for thousands of millennia. Hoards of ancient stone tools found around Olduvai and butchery marks on those bones also tell us how our ancestors were not merely spectators but very much a part of the action.

***

Human life has changed immensely over the millennia, but never so much or so quickly as in the past century. For almost the entire 200,000-year existence of our species, *Homo sapiens*, biology controlled us. We gathered fruits, nuts, and plants; hunted and fished for the animals that were available; and like the wildebeest or zebra, we moved on when resources ran low. Even after the advent of farming and civilization, and the development of cities, we were still very vulnerable to the whims of the weather, and to famine and epidemics.

But in just the past hundred years or so, we have turned the tables and taken control of biology. Smallpox, a virus that killed as many as 300 *million* people in the first part of the twentieth century (far more than in all wars combined) has not merely been tamed but has been eradicated from the planet. Tuberculosis, caused by a bacterium that infected 70–90 percent of all urban residents in the nineteenth century and killed perhaps one in seven Americans, has nearly vanished from the developed world. More than two dozen other vaccines now prevent diseases that once infected, crippled, or killed millions, including polio, measles, and pertussis. Deadly diseases that did not exist in the nineteenth century, such as HIV/AIDS, have been stopped in their tracks by designer drugs.

Food production has been as radically transformed as medicine. While a Roman farmer would have recognized the implements on an American farm in 1900—the plow, hoe, harrow, and rake—he would not be able to fathom the revolution that subsequently transpired. In the course of just one hundred years, an average yield of corn more than quadrupled from about 32 to 145 bushels per acre. Similar gains occurred for wheat, rice, peanuts, potatoes, and other crops. Driven by biology, with the advent of new crop varieties, new livestock breeds, insecticides, herbicides, antibiotics, hormones, fertilizers, and mechanization, the same amount of farmland now feeds a population that is four times larger, but that is accomplished by less than 2 percent of the national labor force compared to more than 40 percent a century ago.

The combined effects of the past century's advances in medicine and agriculture on human biology are enormous: the human population exploded from fewer than 2 billion to more than 7 billion people today. While it took 200,000 years for the human population to reach 1 billion (in 1804), we are now adding another billion people every twelve to fourteen years. And, whereas American men and women born in 1900 had a life expectancy of about forty-six and forty-eight years, respectively, those born in 2000 have expectancies of about seventy-four and eighty years. Compared to rates of change in nature, those greater than 50 percent increases in such a short timespan are astounding.

As Paul Simon put it so catchily, these are the days of miracles.

RULES AND REGULATIONS

Our mastery, our control over plants, animals, and the human body, comes from a still-exploding understanding about the control of life at the molecular level. And the most critical thing we have learned about human life at the molecular level is that *everything is regulated*. What I mean by that sweeping statement is:

- every kind of molecule in the body—from enzymes and hormones to lipids, salts, and other chemicals—is maintained in a specific range; in the blood, for example, some molecules are 10 billion times more abundant than other substances.
- every cell type in the body—red cells, white cells, skin cells, gut cells, and more than 200 other kinds of cell—is produced and maintained in certain numbers; and
- every process in the body—from cell multiplication to sugar metabolism, ovulation to sleep—is governed by a specific substance or set of substances.

Diseases, it turns out, are mostly abnormalities of regulation, where too little or too much of something is made. For example, when the pancreas produces too little insulin, the result is diabetes, or when the bloodstream contains too much "bad" cholesterol, the result can be atherosclerosis and heart attacks. And when cells escape the controls that normally limit their multiplication and number, cancer may form.

To intervene in a disease, we need to know the "rules" of regulation. The task for molecular biologists (a general term I will use for anyone studying life at the molecular level) is to figure out—to borrow some sports terms—the players (molecules) involved in regulating a process and the rules that govern their play. Over the past fifty years or so, we have been learning the rules that govern the body's levels of many different hormones, blood sugar, cholesterol, neurochemicals, stomach acid, histamine, blood pressure, immunity to pathogens, the multiplication of various cell types, and much more. The Nobel Prizes in Physiology or Medicine have been dominated by the many discoverers of the players and rules of regulation.

Pharmacy shelves are now stocked with the practical fruit of this knowledge. Armed with a molecular understanding of regulation, a plethora of medicines has been developed to restore levels of critical molecules or cell types back to normal, healthy ranges. Indeed, the majority of the top fifty pharmaceutical products in the world (which altogether accounted for $187 billion in sales in 2013) owe their existence directly to the revolution in molecular biology.

The tribe of molecular biologists, my tribe, is justifiably proud of their collective contributions to the quantity and quality of human life. And dramatic advances in deciphering information from human genomes are ushering in a new wave of medical breakthroughs by enabling the design of more specific and potent drugs. The revolution in understanding the rules that regulate our biology will continue. One aim of this book is to look back at how that revolution unfolded and to gaze ahead to where it is now heading.

But the molecular realm is not the only domain of life with rules, nor the only branch of biology to have undergone a transformation over the past half-century. Biology's quest is to understand the rules that regulate life on every scale. A parallel, but less conspicuous, revolution has been unfolding as a different tribe of biologists has discovered rules that govern nature on much larger scales. And these rules may have as much or more to do with our future welfare than all the molecular rules we may ever discover.

THE SERENGETI RULES

This second revolution began to flower when a few biologists began asking some simple, seemingly naïve questions: Why is the planet green? Why don't the animals eat all the food? And what happens when certain animals are removed from a place? These questions led to the discovery that, just as there are molecular rules that regulate the numbers of different kinds of molecules and cells in the body, there are ecological rules that regulate the numbers and kinds of animals and plants in a given place.

I will call these ecological rules the "Serengeti Rules," because that is one place where they have been well documented through valiant, long-term studies, and because they determine, for example, how many lions or elephants live on an African savannah. They also help us understand, for example, what happens when lions disappear from their ranges.

But these rules apply much more widely than to the Serengeti, as they have been observed at work around the world and shown to operate in oceans and lakes, as well as on land. (I could just as easily call these the "Lake Erie Rules," but that just seems to lack a sense of majesty). These rules are both surprising and profound: surprising because they explain connections among creatures that are not obvious; profound because these rules determine nature's ability to produce the animals, plants, trees, and clean air and water on which we depend.

However, in contrast to the considerable care and expense we undertake in applying the molecular rules of human biology to medicine, we have done a very poor job in considering and applying these Serengeti Rules in human affairs. Before any drug is approved for human use, it must go through a series of rigorous clinical tests of its efficacy and safety. In addition to measuring a drug's ability to treat a medical condition, these studies monitor whether a drug may cause problematic side effects by interfering with other substances in the body or the regulation of other processes. The criteria for approval pose a high barrier; about 85 percent of candidate medicines fail clinical testing. That high rejection rate reflects, in part, a low tolerance on the part of doctors, patients, companies, and regulatory agencies for side effects that often accompany drugs.

But for most of the twentieth century and across much of the planet, humans have hunted, fished, farmed, forested, and burned whatever and settled wherever we pleased, with no or very little understanding or consideration of the side effects of altering the populations of various species or disturbing their habitats. As our population boomed to 7 billion, the side effects of our success are making disturbing headlines.

For example, the number of lions in the world has plummeted from about 450,000 just fifty years ago to 30,000 today. The King of the Beasts that once roamed all of Africa as well as the Indian subcontinent has disappeared from twenty-six countries. Tanzania now holds 40 percent of all of Africa's lions, with one of their largest remaining strongholds in the Serengeti.

There are similar stories in the oceans. Sharks have prowled the seas for more than 400 million years, but in just the past fifty years, populations of many species around the world have plunged by 90–99 percent. Now, 26 percent of all sharks, including the great hammerhead and whale shark, are at risk of extinction.

Some might say, "So what? We win, they lose. That is how nature works." But that it is not how nature works. Just as human health suffers when the level of some critical component is too low or too high, we now understand from the Serengeti Rules how and why entire ecosystems can get "sick" when the populations of certain members are too low or too high.

There is mounting evidence that global ecosystems are sick, or at least very tired. One measure that ecologists have developed is the total ecological footprint of human activity from growing crops for food and materials, grazing animals, harvesting timber, fishing, infrastructure for housing and power, and burning fuels. Those figures can then be compared with the total production capacity of the planet. The result is one of the most simple but telling graphs I have encountered in the scientific literature (see Figure 2).

Fifty years ago, when the human population was about 3 billion, we were using about 70 percent of the Earth's annual capacity each year. That broke 100 percent by 1980 and stands at about 150 percent now, meaning that we need one and one-half Earths to regenerate

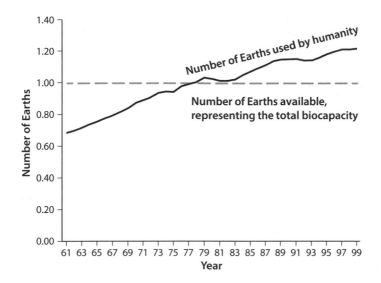

FIGURE 2 The trend in humanity's ecological demands relative to the Earth's production capacity. We are now overshooting what the planet can regenerate by about 50 percent.

Figure from Wackernagel, M., N. B. Schulz, D. Deumling, A. C. Linares et al. (2002) "Tracking the Ecological Overshoot of the Human Economy." *Proceedings of the National Academy of Sciences USA* 99: 9266–9271. © 2002 National Academy of Sciences.

what we use in a year. As the authors of this now annual study note, we have a total of just one Earth available.

We have taken control of biology, but not of ourselves.

RULES TO LIVE BY

As biased it sounds, coming from a biologist, the impact of biology over the past century demonstrates that among all the natural sciences, biology is central to human affairs. There can be no doubt that in facing the challenges of providing food, medicine, water, energy, shelter, and livelihoods to a growing population, biology has a central role to play for the foreseeable future.

Every ecologically knowledgeable biologist I know is deeply concerned about the declining health of the planet and its ability to continue to provide what we need, let alone to support other creatures. Wouldn't it be terribly ironic if, while we race toward and discover more cures to all sorts of molecular and microscopic threats to human life, we continue to just sail on blissfully or willfully ignorant of the state of our common home and the greater threat from disregarding how life works on the larger scale? No doubt most passengers on the Titanic were also more concerned about the dinner menu than the speed and latitude at which they were steaming.

So, for our own sake, let's know all the rules, not just those that pertain to our bodies. Only through wider understanding and application of these ecological rules will we control and have a chance to reverse the side effects we are causing across the globe.

But my goals in this book are to offer much more than some rules, however practical and urgent they are. These rules are the hardearned rewards of the long and still ongoing quest to understand how life works. One of my aims here is to bring that quest to life, as well as the pleasures that come from discovery. My premise is that science is far more enjoyable, understandable, and memorable when we follow scientists all over the world and into the lab, and share their struggles and triumphs. This book is composed entirely of the stories of people who tackled great mysteries and challenges, and accomplished extraordinary things.

As for what they discovered, there is more to gain here than just better operators' manuals for bodies or ecosystems. One of the beliefs that many people have about biology (no doubt the fault of biologists and biology exams) is that understanding life requires command of enormous numbers of facts. Life appears to present, as one biologist put it, "a near infinitude of particulars which have to be sorted out case by case." Another of my aims here is to show that is not the case.

When we ponder the workings of the human body or the scene I encountered on the Serengeti, the details would seem overwhelming, the parts too numerous, and their interactions too complex. The power of the small number of general rules that I will describe is their ability to reduce complex phenomena to a simpler logic of life.

That logic explains, for example, how our cells or bodies "know" to increase or decrease the production of some substance. The same logic explains why a population of elephants on the savanna is increasing or decreasing. So, even though the specific molecular and ecological rules differ, the overall logic is remarkably similar. I believe that understanding this logic greatly enhances one's appreciation for how life works at different levels: from molecules to humans, elephants to ecosystems.

What I hope everyone will find here, then, is fresh insight and inspiration: insight into the wonders of life at different scales; inspiration from the stories of exceptional people who tackled great mysteries and had these brilliant insights, and a few whose extraordinary efforts have changed our world for the better.

***

After five days in the Serengeti, we have seen all of the species of large mammals except one. As we drive back out through the straw-colored grasslands, as if on cue, a novel silhouette appears on the horizon with a prominent telltale horn—a black rhino. With just thirty-one rhinos remaining in the entire Serengeti, it is a rare and thrilling sight. But knowing that there was once more than 1,000 of the animals here, it is also a sober reminder of the challenges ahead. Although, thanks to knowing the molecular rules of human erections, we now have at least five different inexpensive pills that can do the job, rhino horns are still being poached for use as very expensive aphrodisiacs in the Orient.

> *These are the days of miracle and wonder,*
> *And don't cry baby, don't cry*
> *Don't cry*

PART I
EVERYTHING IS REGULATED

CHAPTER 1
THE WISDOM OF THE BODY

The living being is stable. It must be so in order not to be destroyed, dissolved, or disintegrated by the colossal forces, often adverse, which surround it.

—CHARLES RICHET, NOBEL LAUREATE (1913)

The snapping of tree limbs jolted me out of a deep sleep. Peering through the front screen of our large tent, perched on a wooded bluff over the Tarangire River in northern Tanzania, I could not see anything outside in the pitch-black, moonless night. Maybe the wind had toppled a tree? I checked the clock—4 a.m.—and rolled over, hoping to get a couple of more hours of rest.

Then I heard heavy footsteps, crunching at first in front of the tent, then on all sides of us, accompanied by occasional low rumbling, almost purring noises. They were *really close.* My wife Jamie was now awake.

A family of elephants had hiked up the slope from the riverbed to browse on the trees and shrubs on top. With no natural predators, the animals walked wherever they desired, and at 8,000 pounds or more with strong, forklift-like tusks, they simply bulldozed their way through any thicket. As we heard branches and trunks splinter, I wondered about the thin canvas that separated us. With utter disregard for the resting humans nearby and, thankfully, no interest in

FIGURE 1.1 Elephant! Bull moments after a bluff charge, Tarangire National Park.

Photo courtesy of Patrick Carroll.

our rectangular refuges, they munched past dawn before heading back down the hill to drink.

As daylight came, we stepped carefully outside to photograph one straggler. Boy, elephants look even bigger when there is nothing between you and them. This bull was *huge*, more than ten feet tall at his shoulders, with giant ears. Stripping branches and leaves off of small trees, while ignoring the paparazzi peering around the corners of several tents, he seemed content. [Figure 1.1]

Until some noise from a tent spooked him. He trumpeted, pivoted to his left and took some quick steps in our direction.

There is more than one account of what happened next.

In my version, we dashed for the nearest tent, barreled inside, and instantly closed the zipper behind us (because four-ton elephants can't open zippers). We then just stood inside trembling and muttering, trying to regain our composure.

In the biological version of those few seconds, a remarkable number of things happened in my brain and body. Before my mind could even form the thought "Mad elephant! Run!" a primitive part of my brain, the amygdala, was signaling danger to my hypothalamus. This almond-sized command center just above the amygdala promptly sent out electrical and chemical signals to key organs. Through nerves, it signaled the adrenal glands that sit on top of my kidneys to release norepinephrine and epinephrine, also known as adrenaline. These hormones then circulated quickly through the bloodstream to many organs including: my heart, causing it to beat faster; my lungs, to open up airways and increase breathing rate; my skeletal muscles, to increase their contraction; my liver, to release stored sugar for a quick supply of energy; and smooth muscle cells throughout my body, causing blood vessels to constrict, skin hairs to stand on end, and blood to shunt away from the skin, intestine, and kidneys. The hypothalamus also sent a chemical signal, corticotropic releasing factor (CRF), to the nearby pituitary gland that triggered it to release a chemical called adrenocorticotropic hormone (ACTH) that traveled to another part of the adrenal gland and triggered the release of another chemical—cortisol, which increased blood pressure and blood flow to my muscles.

All these physiological changes are part of what is known as the "fight-or-flight" response. Coined and described a century ago by Harvard physiologist Walter Cannon, these responses are aroused by both fear and rage, and quickly prepare the body for conflict or escape. We opted for escape.

SCAREDY CATS

Cannon first became interested in the body's response to fear while conducting pioneering studies on digestion. X-rays had just been discovered when Cannon was a medical student; a professor suggested that he try to use the new gadget to watch the mechanics of the process. In December 1896, Cannon and a fellow student successfully obtained their first images—of a dog swallowing a pearl button. They soon experimented with other animals including a chicken, a goose, a frog, and cats.

One challenge to observing digestion was that soft tissues, such as the stomach and intestines, did not show up well on X-rays. Cannon found that feeding animals food mixed with bismuth salts made their digestive tracts visible, because the element was opaque to the rays. He also explored the use of barium; it was too expensive at the time for research work but was later adopted by radiologists (and still is used in gastroenterology today). In a classic series of studies, Cannon was able to observe for the first time in living, healthy, nonanesthetized animals, as well as in people, how peristaltic contractions move food through the esophagus, stomach, and intestines.

During the course of his experiments, Cannon noticed that when a cat became agitated, the contractions promptly stopped. He jotted in his notebook:

> *Noticed sev times very distinctly (so absol no doubt) that when cat passed from quiet breathing into a rage w struggling, the movements stopped entirely. . . . After about ½ minute the movements started again.*

Cannon repeated the experiment again and again. Every time, the movements resumed once the animal calmed down. The second-year medical student now had another finding to his credit. In what would become the second classic paper of his budding career he wrote, "It has long been common knowledge that violent emotions interfere with the digestive process, but that the gastric motor activities should manifest such extreme sensitiveness to nervous conditions is surprising."

Cannon's knack for experiments soon derailed his plans to become a practicing physician. His talent, rigor, and work ethic so impressed the distinguished faculty of the Department of Physiology at Harvard that he was offered an instructorship on graduation.

THE NERVOUS STOMACH

In his own laboratory, Cannon aimed to figure out how emotions affected digestion. He observed that emotional distress also ceased digestion in rabbits, dogs, and guinea pigs, and from the medical literature that also seemed to be true of humans. The connection between emotions and digestion suggested some direct role of the nervous system in controlling the digestive organs.

Cannon knew that all the outward signs of emotional stress—the pallor caused by the contraction of blood vessels, "cold" sweat, dry mouth, dilation of pupils, skin hair standing on end—occurred in structures that are supplied by smooth muscle and innervated by the so-called sympathetic nervous system. The sympathetic system comprises a series of neurons that originate from the thoracic-lumbar region of the spinal cord and travel out to clusters of nerve cells (called ganglia). From there, a second set of generally much longer neurons extend to and innervate target organs. Most of the body's organs and glands receive sympathetic input, including the skin, arteries, and arterioles, the iris of the eyes, the heart, and the digestive organs. These same organs also receive input from nerves originating in the cranial or sacral parts of the spinal cord. [Figure 1.2]

To figure out what stopped the activity of the stomach and intestines under emotional stress, Cannon and his students conducted a series of simple but fundamental studies. One approach was to sever the nerves leading to the digestive organs. Cannon found that when the vagus nerve (originating in the cranial system) was severed but the splanchnic nerve (part of the sympathetic system) was left intact, the inhibition of peristalsis could still be induced by fear. In contrast, when the splanchnic nerves were cut and vagus remained intact, there was no response to fear. These results showed that the inhibition of peristalsis induced by emotion required the sympathetic splanchnic nerves.

Cannon had noticed that the inhibition of gastric activity often long outlasted the presence of whatever provoked the response. This suggested to him that there might be a second mechanism beyond direct nervous impulses that might prolong the agitated state. It had been reported that adrenalin, a substance extracted from the central portion of the adrenal glands, when injected into the bloodstream could produce some of the effects produced by stimulation of the sympathetic nervous system. Cannon wondered whether the adrenal glands might be involved in the body's response to fear and anger.

To test this possibility, Cannon "made use of the natural enmity" between dogs and cats. He and a young physician, Daniel de la Paz,

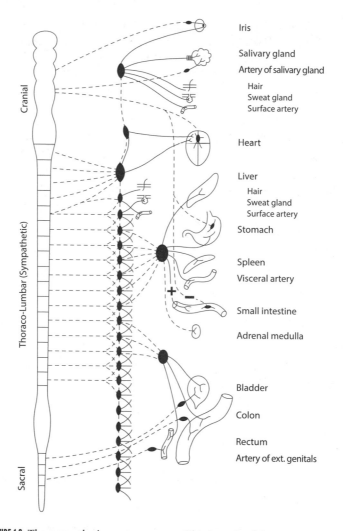

FIGURE 1.2 The sympathetic nervous system. This branch of the autonomic nervous system connects to various glands and smooth muscles to maintain homeostasis and to mediate the flight-or-fight response. Nerves emanating from the cranial and sacral regions generally act in opposition to those emanating from the thoraco-lumbar region (trace, for example, the innervation of the small intestine).

Figure adapted from *The Wisdom of the Body* by Walter B. Cannon (1963), modified by Leanne Olds.

compared blood samples of cats taken before and after they had been exposed to the stress of barking dogs. They discovered that the blood of frightened cats contained a substance that when applied to a small strip of isolated intestinal muscle stopped it from contracting. This was the same effect observed when adrenalin was applied to the muscle strip.

Epinephrine was one of the components of "adrenalin" produced by the adrenal glands. Cannon and his colleagues also found that epinephrine sped up heart rate, the release of sugar from the liver, and even blood clotting. These same effects were triggered by pain, as well as by fear or anger. None of these effects occurred when the adrenal glands were removed, or when the nerves leading to the adrenal glands were cut. Thus, the sympathetic nervous system and adrenal glands worked in concert to modulate other body organs in stressful conditions.

Cannon suggested that the responses induced by epinephrine reflected the "emergency" function of the adrenal glands in preparation for fight or flight, or in response to pain. A firm adherent to Darwin's principle of natural selection, Cannon interpreted the roles of the adrenal system through that lens:

> The organism which . . . can best muster its energies, can best call forth sugar to supply the laboring muscles, can best lessen fatigue, and best send blood to the parts essential in the run or fight of its life, is most likely to survive.

Cannon's student Philip Bard subsequently demonstrated that the hypothalamus is the critical part of the brain for control of the so-called involuntary (autonomic) functions of the nervous system, including digestion, heart rate, respiration, and the fight-or-flight response. Both this part of the brain and these emergency responses are ancient. This same set of responses helped our ancestors avoid lions and hyenas on the savannah, just as they help pedestrians to dodge taxis in New York today, or tourists to run from elephants.

A SCIENTIST-SOLDIER

Cannon was an Ivy League but not an Ivory Tower scientist. In 1916, three years into World War I, as the battlefield in Europe turned into a horrific stalemate that produced enormous casualties, it seemed increasingly likely that the United States might be drawn into the conflict. Cannon was asked to chair a special committee of physiologists to advise the government on ways to protect the lives of soldiers and civilians. He learned that one of the most serious problems in battlefield medicine was the development of shock in wounded soldiers. Cannon recognized some of the shock symptoms—rapid pulse, dilated pupils, heavy sweating—from those he had observed in his experimental studies of animals under stress. Wounded soldiers who exhibited these symptoms often went downhill quickly and died. "Are there not untried ways of treating it?" he asked a fellow physiologist.

Cannon was so taken with the problem of shock that he began some animal experiments to see whether he could figure out ways to mitigate the syndrome. When the United States did finally enter World War I in April 1917, Cannon was forty-five years old and the father of five, and could have easily been excused from service. Instead, he volunteered as a member of a Harvard Hospital Unit that was one of the first American medical teams to go to Europe. Cannon requested to serve in a shock ward near the front lines in northern France.

Cannon said goodbye to his family in Boston, took a train to New York, and boarded the troopship *Saxonia* bound for England. The voyage overseas would take eleven days. To avoid detection by German submarines, the ship was blacked out at night, with all of its portholes closed. While ships usually have lights at both ends so as to avoid collisions, the *Saxonia* lit only its stern, to help draw any torpedo off target. Eight days into the voyage, as the ship drew nearer to the English coast, the orders came to sleep in one's clothes; if hit, it was better to jump into the lifeboats fully dressed. As the ship hit choppy seas in rain and fog, Cannon was relieved, "not a favorable condition, I should say, for good hunting," he wrote to his wife Cornelia. The appearance of a British destroyer escort further eased anxieties.

After arriving safely in England, Cannon continued on to the first of several field hospitals. A wave of casualties soon arrived from a major British offensive. Although Cannon had not practiced any medicine since his graduation from medical school seventeen years earlier, he asked to assist in the operating room, dressed wounds, and worked in the wards.

Cannon then moved to a hospital nearer to the front. He watched helplessly the heartbreaking, rapid decline of scores of soldiers. Why the soldiers died was a mystery that Cannon and several other American and British physiologists were hell-bent to solve.

One important clue to shock came from the then-novel approach of measuring soldiers' blood pressures, not just their pulses. Healthy soldiers had pressures of about 120–140 (mmHg; the abbreviation stands for millimeters of mercury), while shock patients had pressures below 90. It was learned that if this fell to 50–60, the patient did not recover.

A low blood pressure meant that vital organs would have difficulty obtaining sufficient fuel and disposing of waste. Early in his time in France, Cannon decided to measure the concentration of bicarbonate ions in the bloodstream of shock patients, a critical component of the blood's buffering system. He discovered that the patients had lower levels of bicarbonate, which meant that the normally slightly alkaline blood had become more acidic. And he found that the more acidic the blood was, the lower the blood pressure and the more severe the shock were. Cannon proposed a simple possible therapy: administer sodium bicarbonate to shock victims.

Cannon reported the first results in a letter to his wife Cornelia in late July 1917, just two months after his arrival in Europe:

> Well, on Monday there was a patient with a blood pressure of 64 (the normal is about 120) millimetres of mercury and in a bad state. We gave him soda [sodium bicarbonate], a teaspoonful every two hours and the next morning the pressure was 130. And on Wednesday a fellow came in with his whole upper arm in a pulp . . . such cases usually die. At the end of the operation he had the incredibly low pressure of 50; soda was started at once and the next morning the pressure was 112.

Cannon described three other soldiers who had been treated that same week and also had been "snatched from death," including one who was given the sodium bicarbonate intravenously and whose rapid respiration and pulse eased quickly.

Cannon and the Allied medical command were thrilled by this innovation. Since shock was often brought on by surgery, the use of bicarbonate was adopted as a standard preventative measure in all critical cases. Cannon and his colleagues also advocated other procedures for warding off the development of shock, including protecting wounded soldiers from exposure by wrapping them in warm blankets, giving warm fluids, transporting on dry stretchers, and using lighter forms of anesthesia during surgery.

To promote these methods, Cannon organized the training and deployment of "shock teams" to treat shocked soldiers on or near the battlefield. To see how the teams performed in battle, he went on an inspection tour close to the front.

In mid-July 1918, he was visiting a hospital near Chalons-sur-Marne, in eastern France. After spending an evening socializing with other doctors, Cannon retired to bed. He could hear guns firing in the distance, but that was typical. Just before midnight, Cannon was jolted awake by "the most stupendous, the most terrific, the most inconceivably awful roar . . . like thousands of huge motor trucks rushing over cobblestones." He jumped to his window and saw entire horizon lit up with gunflashes and shellbursts. He heard the zip-sish sound of a shell passing nearby, which exploded near the hospital. Shells continue to hit within a mile of the building, one about every three minutes for four straight hours.

In the middle of the massive German assault, Cannon was called to the shock ward as the first few casualties were brought in. Then came a flood of wounded—eventually more than 1,100 would arrive that day. As the shock ward filled, Cannon heard a deafening crash—a shell struck the next ward, just twenty feet away, blowing off the roof and sending shrapnel through the walls of his ward. Dust, smoke, and gasses from the explosion filled the air, but Cannon and the rest of the teams stayed at their stations until all patients had been attended to and moved to safer quarters behind the front.

The battle ended up being a turning point in the war. The German drive stalled, and the Allies pushed eastward over the following weeks and months. Cannon followed the leading front into formerly German-held territory. He saw French towns in complete ruins, desolate landscapes denuded of all greenery, and longs columns of enemy prisoners. At last, the streams of Allied wounded coming into the hospitals slowed to a trickle, then stopped altogether; the war was over. Cannon wrote to his wife, "There is satisfaction now in knowing . . . that we were serving the wounded close to the center of the struggle which changed the whole history of the world."

Cannon's exemplary performance during the war was recognized by a series of promotions. In a span of just fourteen months, he went from first lieutenant to captain, then to major, and finally to lieutenant colonel. He was awarded the Order of the Bath by the British and cited by General Pershing, the leader of the American forces in Europe: *"For exceptional meritorious and conspicuous services as instructor in shock treatment."* After a joyous celebration in Paris, he sailed back home to the United States, his wife, children, and Harvard laboratory in January 1919. [Figure 1.3]

THE WISDOM OF THE BODY

Cannon's experiences in France had a profound impact on the physiologist. They gave him a poignant, first-hand understanding of the important parameters for the maintenance of human life. Combined with his knowledge of the control of digestion, respiration, heart rate, and the responses to stress in animals, Cannon was provoked to think about the body's ability to react to disturbances and yet to maintain critical functions within fairly narrow ranges.

To Cannon, it appeared that many activities of the nervous and endocrine systems served to prevent wide oscillations and to hold the internal conditions of the body—temperature, acidity, water, salts, oxygen, and sugar—fairly constant. He knew too well that if these narrow limits are breached, serious illness or death often follows. For example, blood pH, a measure of acidity, is maintained near 7.4; if it drops to 6.95, coma and death result, and if it rises to 7.7, convulsions and seizures occur. Similarly, calcium levels are maintained around 10 milligrams per 100 milliliters of blood; half that level causes convulsions, double that level causes death.

Cannon began to speak of the innate "wisdom of the body" in lectures and papers. "Our bodies are built to take very effective care of themselves, in many ways which we have become aware of only in

FIGURE 1.3 Walter B. Cannon in his Army uniform.

recent years," Cannon wrote. One of the recent breakthroughs was in understanding the role of insulin in controlling blood sugar. Cannon noted how when sugar levels rise after a meal, the vagus nerves stimulate the pancreas to secrete insulin, which causes excess sugar to be stored. Conversely, if sugar levels fall, other nerves in the autonomic system trigger the adrenal glands to liberate sugar from the liver. In this way, Cannon said, "the organism automatically restricts the range over which the percentage sugar in the blood may shift."

Cannon emphasized how most organs received dual nervous inputs that, as a rule, opposed each other. With this wiring, organ activity can be increased or decreased depending on conditions. Impressed by the body's ability to adjust to disturbances, Cannon coined a new term to describe the steady states maintained in the body: *homeostasis* (from the Greek *homeo* meaning "similar" and *stasis* meaning "standing still"). This was not some lofty or abstract philosophical notion; Cannon's concept was firmly rooted in three decades of physiological research. Homeostasis was fundamentally a matter of regulation. That is, physiological processes existed in the body that operated to maintain—to regulate—body conditions within certain ranges.

Cannon first elaborated his ideas in the scientific literature, then in a popular science book titled *The Wisdom of the Body*. He offered several lines of evidence to support his claim that stability was due to active regulation. First, he stressed that the constancy exhibited by body functions in the face of all sorts of external disturbances and variables indicated the presence of regulatory mechanisms that maintained steady states. Second, he argued that states remain steady because factors exist that resist change in either direction, positive or negative. Third, he pointed out that there was substantial evidence that multiple cooperating factors often act either simultaneously or in succession to maintain a state, such as the acid-base balance of the blood. And fourth, he suggested that the existence of some regulatory factor acting in one particular direction implied the existence of factors acting in the opposite direction, as had been shown for blood sugar.

In short, Cannon asserted that everything in the body is regulated. And so he concluded, "regulation in the organism is the central problem of physiology."

Grounded in Cannon's body of work on digestion, thirst, hunger, fear, pain, shock, and the nervous and endocrine systems, and made accessible by his lucid writing, homeostasis became a fundamental concept in physiology and biology. Some compared it to Darwin's principle of natural selection as one of the seminal integrative ideas in biology.

Cannon believed that the implications of homeostatic mechanisms to medicine were far reaching and very positive. He shared his "Reasons for Optimism in the Care of the Sick" in an address to Boston-area physicians that was subsequently published in the *New England Journal of Medicine*. He began his presentation with typical modesty:

> That you, a group of physicians who are daily confronting the practical problems of sick men and women, should ask me, a physiologist, a laboratory recluse, to address you is a surprising fact. Perhaps my presence here calls for some explanations from you—and for some apologies from me! . . . All that I propose to do as a physiologist is to draw forth some suggestions from years of research and reading and thinking about the workings of the organism . . . that may be useful as laying a basis for optimism in medical practice.

Cannon then recounted how when some factors

> tip the organism in one direction or the other, internal adjustments have promptly been called into service which have prevented the disturbances from going too far and have tipped the organism back to its normal position. Note that these are not processes which [sic] we manage ourselves. They are automatic adjustments.

In light of these marvelous powers of self-regulation Cannon asked, "If the body can largely care for itself what is the function of the physician?"

He explained that doctors' services are called for when these mechanisms are overwhelmed or malfunctioning. Cannon emphasized how many of the newer therapies available to physicians—insulin, thyroxin, antitoxins—were natural components of the body's self-regulatory system. The physician's role was thus to reinforce or to restore the natural homeostatic mechanisms of the body. Cannon suggested that the power of these mechanisms, and the increasing ability of physicians to bolster them, were cause for optimism in medicine.

Cannon had the powerful ideas that regulation is the central matter of physiology, and that abnormal regulation is the central issue of medicine. Coincidentally, at the very same time that Cannon was expressing these pivotal ideas, another biologist was reaching the conclusion that regulation was the central issue in nature on a much larger scale.

CHAPTER 2

THE ECONOMY OF NATURE

[T]he study of the regulation of animal numbers
forms about half the subject of ecology, although it
has hitherto been almost untouched.

—CHARLES ELTON

The charge was a bluff. The elephant took only a few steps, just enough to let us know which mammal was boss of that hilltop.

Once our heartbeats returned to normal, and after he worked his way down the slope, we ventured back outside to survey the aftermath of the night's raid. There was a wake of broken trees, naked branches, and the lingering aroma of dung (theirs, not ours). Elephants are prodigious producers of the latter: their hundred-foot-long intestines manufacture up to 200 pounds of manure per day to keep up with the more than 200 pounds of food and fifty gallons of water they consume.

Given the absence of any natural predators, and their enormous appetites, one might ask why East Africa is not overrun with, or stripped bare by, elephants? Perhaps it is because African elephants, the largest of all land animals, reproduce very slowly? Females do not mature until their teens, they give birth to just a few young in their lifetimes, and it takes twenty-two months of gestation for a 250-pound baby elephant to develop.

It was Charles Darwin who famously dispelled that explanation in *On the Origin of Species*:

> The elephant is reckoned the slowest breeder of all known animals, and I have taken some pains to estimate its probable minimum rate of natural increase; it will be safest to assume that it begins breeding when thirty years old, and goes on breeding till ninety years old, bringing forth six young in the interval, and surviving till one hundred years old; if this be so, after a period of from 740 to 750 years there would be nearly nineteen million elephants alive, descended from the first pair. Indeed, in fewer than 50 generations or about 2500 years, the total *volume* of elephants would exceed that of the planet.

That figure is ridiculous. But consider the other end of the size spectrum. A typical bacterium, such as the *Escherichia coli* that populates our intestines, weighs about one-trillionth of a gram (one trillion bacteria weigh one gram; an elephant weighs about four million grams). Based on a maximum doubling time of twenty minutes, one can calculate how long it would take for one bacterium to give rise to enough bacteria to equal the weight of the Earth. The answer: just two days.

But the world is not made of solid elephants or bacteria.

Why? Because there are limits to the growth and numbers of all creatures.

Darwin recognized that. And he understood it because Reverend Thomas Malthus stated it long before in his landmark *Essay on the Principle of Populations* (1798):

> Population, when unchecked, increases in a geometrical ratio. . . . The germs of existence contained in this spot of earth, with ample food, and ample room to expand in, would fill millions of worlds in the course of a few thousand years. Necessity, that imperious all pervading law of nature, restrains them within the prescribed bounds. The race of plants and the race of animals shrink under this great restrictive law.

But just how are these "bounds" set? And how are they set differently for different creatures? Darwin did not know. These questions were not pursued in earnest until another young English naturalist went on an expedition to some other remote islands, also encountered an odd assortment of creatures, became gripped by a mystery, had an epiphany (or several) during his adventures, wrote a great book, and founded a new field of science. Charles Elton is nowhere near as famous as Darwin or Malthus, but he is known to biologists as the founder of modern ecology, and the central mystery that gripped him was how the numbers of animals are regulated.

A JOURNEY TO THE ARCTIC

The *Terningen* pitched and rolled on the heaving, icy Barents Sea, and twenty-one-year-old Oxford University zoology student Charles Elton heaved right along with it. The two-masted schooner had left Tromsø two days earlier under the June midnight sun, bound for Bear Island, a desolate, rocky island well above the Arctic Circle, whose most prominent feature was aptly named Mount Misery.

Elton was part of a small advance party of the first so-called Oxford University Expedition to Spitsbergen, a team of twenty students and faculty from various disciplines—ornithology, botany, geology, and zoology—who aimed to carry out an extensive geographic and biological survey of the largest island in the Arctic archipelago northwest of mainland Norway. It was a bold and ambitious undertaking to venture far into storm-tossed and ice-strewn waters, to go ashore and traverse largely uninhabited and partly snow- and ice-covered islands, while risking the whims of some of the most intense weather on the planet; not to mention the fact that none of the team had any prior experience in the Arctic. But it was just the sort of adventure, and personal test, that the select group of Oxford men sought.

The voyage to Bear Island was a bit less than 300 miles and a severe test for Elton, who had never even left England before. The ship was a converted sealing boat whose sleeping quarters previously served to hold blubber, the smell of which could not be removed. That combined with the rough seas spelled utter misery.

From an early age, Elton had been captivated by wildlife. He spent countless days walking through the English countryside, watching birds, catching insects, collecting pond animals, and examining

FIGURE 2.1 The First Oxford University Spitsbergen Expedition team, 1921. Charles Elton is fifth from the right in the turtleneck sweater, Vincent Summerhayes is to the immediate left of Elton.

Photo from account by C. S. Elton written in 1978–1983. Courtesy of Norsk Polarinstitutts Library, Tromsö, Norway.

flowers. He began a diary of all of his wanderings and observations at age thirteen. Elton had been invited on the expedition by his tutor, the eminent zoologist and writer Julian Huxley (grandson of close Darwin ally Thomas Huxley), who was impressed with his student's passion for and command of natural history. He had relatively little time to prepare, as his slot on the expedition was confirmed just one month before sailing. His father provided some money, his brother lent him his Army boots and other gear, and his mother helped him put together clothes for the two-month voyage. Elton was, by his own account "very inexperienced, very raw indeed," and had only previously done some rough camping. Huxley encouraged Elton and reassured his parents, "Please assume that there is no danger in the expedition further than that involved, say, in elementary Swiss climbing, certainly much less than difficult mountaineering and indeed negligible."

By just the third day of the voyage, that pledge had crumbled. Forty-six-year-old medical officer and accomplished mountaineer

Tom Longstaff was the most seasoned voyager aboard. Seeing Elton suffering, he said, "I must do something for that poor boy," and proceeded to medicate him with large doses of brandy. With nothing in his stomach, Elton was so drunk when the landing party of seven finally went ashore at Walrus Bay on the island's southeast coast that he arrived sitting on top of a load of baggage in the whale boat, singing at the top of his lungs.

Once sober, Elton set up camp with his companions inside the ruins of an old whaling station near the shore, where the ground outside was littered with whale bones, walrus skeletons, a polar bear's skull, and a couple of arctic fox skulls. The plan was for the party to spend a week exploring the southern portion of the twelve-mile-long island before rejoining the full team and sailing on to Spitsbergen, about 120 miles to the north. The four ornithologists in the group were to focus on the birdlife, while Elton and botanist Vincent Summerhayes would survey all plants and other animals.

The birders were a bit zealous about their mission, especially F.C.R. Jourdain, the leader of the expedition. At 2:30 a.m. of the first "night" on shore (there were twenty-four hours of daylight), Jourdain awakened Elton, Summerhayes, and Longstaff with the command to get up and get to work, as there was no time to lose. Longstaff replied, "I must have my eight hours sleep," and the veteran explorer whispered to Elton and Summerhayes, "Don't take any notice." Elton soon appreciated Longstaff's wisdom, as the birders were completely exhausted by evening. Elton learned to pay close attention to Longstaff's sage advice whenever offered.

After a good night's rest, Elton and Summerhayes went out to explore and collect. Elton dreamed of one day knowing what animals "are doing behind the curtain of cover." Now he would have a chance to peer into an alien world that few had ever visited, let alone studied.

Situated where the cold polar current meets the Gulf Stream from the west, the island was constantly shrouded in fog, when it was not pounded by sleet or blizzards. Largely flat and dotted with dozens of lakes throughout its interior, and marked by a barren almost moon-like landscape in the north, the island was not without its wonders. On the southern end stood several tall cliffs overlooking the sea, populated by hundreds of thousands of seabirds. Among the most numerous were black and white Common and Brunnich's guillemots, black-legged kittiwakes, and gray fulmar petrels; Elton also saw little auks, Norwegian puffins, and glaucous gulls.

Elton and Summerhayes started their surveys near the camp, then eventually fanned out to various lakes and other features. Elton quickly discovered that he had an excellent, determined partner in the small, slightly-built botanist. Just three years older, Summerhayes had seen more of the world, including fighting in the battle of the Somme (1916). Summerhayes was as keen to identify all the Arctic plants as Elton was to discover the animals, and in the course of their wanderings he taught Elton a great deal of botany.

Despite the weather and the shortness of time, the two men were able to survey all the island's various habitats for plant and animal life. The task was made easier by the relative sparsity of species. For instance, throughout the island, there were no butterflies, moths, beetles, ants, bees, or wasps whatsoever; the insect life consisted mostly of flies and primitive bugs known as springtails.

Elton had brought various means for collecting and preserving such treasures. He caught flying insects in a butterfly net, and shook other bugs out of plants and leaves onto a white sheet, or uncovered them by flipping over stones. The insects were killed with cyanide, and Elton put them inside tissue paper with labels and packed them into cigar boxes. For aquatic creatures Elton used a series of nets with different mesh sizes. He preserved some aquatic animals inside glass tubes containing alcohol or formalin, which he then corked and sealed with wax.

Elton paid particular notice to what enabled each species to survive in the generally barren and harsh conditions. For example, it was obvious that the seabirds lived off the sea and that the vast quantities of manure they dropped from the cliffs fertilized the lush vegetation growing below. Other birds Elton found inland, such as the snow bunting, dined on sawflies, while the arctic skua lived off other birds, stealing their food and eggs.

As the days wore on and food supplies began to dwindle, the skua weren't the only creatures living off the birds. The ornithologists collected many eggs and birds. The egg shells were preserved by blowing out their contents (Jourdain once blew so many that he fainted), which were then made into omelettes. And after skinning, the bird carcasses were put in cooking pots. Elton and the team were surprised at how many species were edible, and he jovially wrote home about having guillemot eggs for breakfast and "stew of fulmar, eider duck, long-tailed duck, purple sandpiper, snow bunting, vegetables & rice!"

Toward the end of their planned week on the island, a gale blew in that would make it impossible for the ship to approach the island safely. Longstaff realized that they would have to stay longer and needed to get a message to the *Terningen*, which was waiting in Tromsø. He asked Elton to walk with him to the northeastern corner of the island (an expeditionary rule was to always go out in pairs for safety) to a coal mine with a wireless station. It was a brutal seven-mile, four-hour hike in driving snow and sleet to reach the mine, and then another four hours back again across knife-edge stones and slushy ground.

Despite the storms, running low on bread, being out of margarine for cooking, and having nearly worn through his Army boots in just a week, Elton declared that life on Bear Island "is rather fun."

After four days of battling strong gales, the ship was able to return to Bear Island and to retrieve Elton and his comrades just a few days behind schedule. The entire expedition team then headed farther north toward Spitsbergen—and straight into another gale that pushed "growlers" (chunks of dangerous, greenish sea ice) into their path.

Once that storm subsided, the *Terningen* made its way up the west coast of Spitsbergen. About halfway up the 280-mile-long island, the ship turned toward the azure waters of Ice Fjord, and the clouds parted. The team was treated to a stunning panorama of jagged, snow-capped mountain peaks and shimmering glaciers that crept down pure white valleys to the sea. A whale's spout and then a pod of porpoises broke the glassy surface, while strings of petrels and auks darted just above the water.

More than 200 miles wide, Spitsbergen presented a vast opportunity, and challenge, for exploration. Elton landed first on a smaller island west of the main island called Prince Charles Foreland, where he would spend about ten days surveying with Summerhayes. Longstaff helped Elton establish a campsite on the scrabble near a large, several-mile-long lagoon that was half covered with ice, on which many seals lounged. Before returning to the ship, Longstaff cautioned

Elton not to walk across the lagoon ice. Elton had learned that Longstaff only gave advice once, and if one forgot it, well, that was one's own mistake.

After nine fruitful days on the island collecting and observing the animals, Elton made that mistake. He was out walking with geologist R. W. Segnit on the ice attached to the shoreline of the lagoon. He stepped on a weak spot created by runoff from the shore and broke through into the frigid water. He was saved from going under only by his ruck-sack, which held him up a bit, and was able to climb carefully back up on the ice. However, freezing cold, momentarily stunned, and wearing sun-goggles, Elton did not realize that he had been turned around and almost stepped right back into the hole. Only a shout from Segnit stopped him from another misstep that would have likely drowned him. Shivering with hypothermia, Elton was able to return to camp to dry out; he left the island two days later for a longer stay at Klaas Billen Bay in the interior of the main island.

FIGURE 2.2 Charles Elton studying pond life at Bruce City, Spitsbergen, 1921.

Photo from account by C. S. Elton written in 1978–1983.
Courtesy of Norsk Polarinstitutts Library, Tromsö, Norway.

Despite being farther north, and more snow- and ice-covered than Bear Island, the summer temperatures on Spitsbergen were much warmer (some days over fifty degrees Fahrenheit) and more comfortable, at least when it was not snowing. Elton took advantage of the conditions to conduct some actual experiments on the adaptations of Arctic animals. To learn more about how they survived in the climate and which habitats they could live in, he tested the tolerance of crustaceans and their eggs to freezing and thawing, and to different concentrations of sea water. [Figure 2.2]

Elton found very similar birdlife, vegetation, and insect life on Spitsbergen as on Bear Island but saw more evidence of larger mammals, including two species of arctic foxes, reindeer (in the form of shed horns), and seals. Then, one morning while the breakfast bacon was frying, a shout came up from below the campsite: "Polar bear!"

No one knew how the unwelcome visitor had reached the bay over the glaciers and valleys. But now that it had found their camp, there was sadly no alternative but to shoot it. Elton, ever the zoologist, did a very thorough inspection of the body, both outside and in, and was disappointed not to find any hitchhiking parasites for his collections.

After more than two months of collecting and living on the tundra, the now-veteran Arctic explorer assembled his thirty-three boxes and bundles of gear and specimens, loaded them all onto the ship himself, and sailed home.

UP AND DOWN AN ARCTIC FOOD CHAIN

Once safely back at Oxford, Elton and Summerhayes assembled the data they had gathered from their surveys. Many naturalists, especially those with a collector's bent, might have been disappointed at the meager pickings on the Arctic islands. But Elton realized that the relative scarcity of species presented a rare opportunity to describe all the transactions and relationships among an entire community of organisms—to peel back the curtain on animal life.

While previous naturalists had seen a community as one entity, or as a collection of species, Elton took a novel, functional approach. It was obvious to Elton that in the economy of Arctic islands, the precious commodity was food. So he traced where every creature's food came from.

Food was extremely scarce on land, but it was plentiful in the sea, so he started there. He knew that marine animals (plankton, fish) were eaten by the seabirds and by the seals. The seabirds were eaten in turn by the arctic fox (as well as by the skua and glaucous gull), while the seals were eaten by polar bears. These relationships formed what he dubbed "food chains."

But the connections among the inhabitants of the tundra extended far beyond a few animals. The droppings from the seabirds

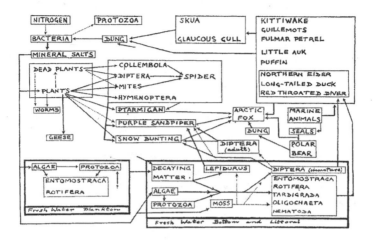

FIGURE 2.3 The food chain on Bear Island. The first food web drawn by Charles Elton. To follow the food chain on land, start with the nitrogen and bacteria at upper left and trace the links all the way to the arctic fox.

From Summerhayes and Elton (1923).

contained nitrogen, which was used by bacteria, which nourished plants, which produced food for insects, both of which were consumed by land birds (ptarmigan, sandpiper), which in turn were food for the arctic fox. In this manner, the food chains in a community were connected into larger networks that Elton dubbed "food-cycles," later called food "webs." Elton drew a schematic of these chains and webs, the first of its kind, in a paper published with Summerhayes in 1923. [Figure 2.3]

LEMMINGS AND LYNX

Elton himself moved quickly up the academic food chain, as well as the ranks of the Oxford University Expeditions. After receiving his undergraduate degree in 1922, he was appointed to a departmental lectureship in 1923, and named chief scientist for a new expedition to Spitsbergen.

It was extraordinary responsibility for a 23-year-old, but it was par for the Oxford enterprise. Polar exploration was largely a young man's game, and Elton was in what proved to be remarkable company, as many expedition members subsequently distinguished themselves in other endeavors or perished trying. The organizer of the first expedition was George Binney, who was just a twenty-year-old undergraduate at the time. Binney would also organize the 1923 expedition and a much more ambitious and complex 1924 expedition. He was later knighted for his exploits during World War II, in which he organized operations to run through the German blockade of Sweden. The 1923 expedition added twenty-one-year-old Sandy Irvine, who would attempt to climb Mount Everest the next year and disappeared with George Mallory just a few hundred meters from its summit. Medical officer Longstaff was the first person to summit a mountain over 7,000 meters and would later climb peaks all over the world. The medical officer on the 1924 expedition, and Elton's sometimes tent-mate, was twenty-five-year-old Australian Rhodes Scholar Howard Florey, who would develop penicillin into a drug during World War II and share the 1945 Nobel Prize in Physiology or Medicine.

The 1923 expedition had a new make-up and destination. Binney sought a leaner, more efficient enterprise and trimmed the team to fourteen: seven scientists and seven "henchmen" who were able hunters, rowers, and mechanics. The *Terningen* steered this time for Spitsbergen's little-explored sister island North East Land (Nordaustlandet). Despite its proximity to Spitsbergen, the island proved much less accessible. The ship was thwarted by thick belts of ice that surrounded the island and broke its propeller while trying to force its way through a strait. Partially crippled, the ship was pushed about by the ice floes, and the plans to explore the island had to be abandoned.

But for Elton, the journey was not a bust. On the way back to England, the ship stopped as usual at Tromsø. Elton went browsing in a bookstore and happened across a large tome on Norwegian mammals titled *Norges Pattedyr* by a Robert Collett. Although Elton did not read Norwegian, he was intrigued enough to plunk down one of the remaining three British pounds he had left in his pocket for the journey home for a book that Elton later declared "changed my whole life."

Back in Oxford, Elton obtained a Norwegian dictionary and laboriously worked out a word-by-word translation of one part of the book. There were fifty or so pages on lemmings that, although he had never seen one of the small guinea pig–like animals, enthralled Elton. He was captivated by Collett's descriptions of so-called lemming years, when the rodents swarmed out of Scandinavian mountainsides and tundra in such massive numbers in the autumn that citizens had taken note of the extraordinary phenomenon for centuries.

Elton drew up charts of the reported events and found that they occurred with a fairly regular periodicity of three to four years. He made maps and noticed that migrations involving different lemming species in different parts of Scandinavia seemed to occur mostly in the same years. He stared for hours at the maps spread around the floor of his cubicle in an old building at Oxford, thinking there must be something important he was missing. Then, sitting on a seat in the lavatory, it came to him "in a flash." Like Archimedes in his bathtub, Elton on his toilet got the idea that the lemmings were the "overflowing" of a periodically increasing population. At the time, zoologists had assumed that animal numbers largely remained steady. But Elton now realized that they could fluctuate dramatically.

Digging further, Elton wondered how general the phenomenon was. Even though there were no direct reports of lemming migrations in Canada, Elton was able to infer those events by thinking about food chains. He had previously read a book by a Canadian naturalist that described fluctuations in the populations of other mammals, including the arctic fox. Knowing that the foxes eat Canadian lemmings, Elton located a table of the number of fox skins taken by the fur-trading Hudson Bay Company and, sure enough, found a good correlation between a spike in the number of fox skins and the timing of lemming years in Norway.

The implications for Elton's food chain concept multiplied. Lemmings were also preyed on by birds. Elton noted that short-eared owls gathered in numbers in lemming years in southern Norway, as did peregrine falcons that preyed on the owls.

Lemmings were not the only prey whose numbers fluctuated wildly. Elton learned that the populations of Canadian rabbit (or "varying hare") also oscillated, spiking enormously about every ten

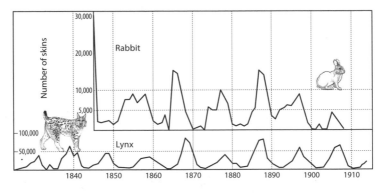

FIGURE 2.4 The ten-year cycles of lynx and rabbits in the Canadian north. Elton studied the records of the Hudson Bay company to determine that the populations of lynx and rabbits surge and crash with a ten-year periodicity.

From Elton (1924).

years before plummeting. The rabbits are the favorite prey of the Canadian lynx. "It lives on Rabbits, follows the Rabbits, thinks Rabbits, tastes like Rabbits, increases with them, and on their failure dies of starvation in the unrabbited woods," wrote one naturalist. The cat's fur meanwhile was the favorite quarry of fur trappers for the Hudson Bay Company. The company, it turned out, kept careful records of furs taken every year from about 1821 on. When plotted, the number of furs taken also showed a striking ten-year cycle that correlated with the rabbit cycle. [Figure 2.4]

Elton thought these cycles offered valuable glimpses into the workings of animal communities. They demonstrated the remarkable power of populations to increase dramatically, particularly if unchecked. By the same token, the abrupt crash in lemmings and rabbits showed that there were also forces, such as epidemic disease, that could rapidly eliminate a large proportion of animals. And these cycles showed how the numbers of one species can influence the numbers of other species through food chains.

In short, the cycles demonstrated that the numbers of animals are regulated in a variety of ways. Elton documented these "Periodic Fluctuations in Numbers of Animals" in a forty-five-page paper he crafted in 1924. He did not know it at the time, but he was laying a cornerstone of the new science of ecology. The young naturalist would next lay the foundation.

LIFE IS ALL ABOUT FOOD

In 1926, Elton's former tutor Julian Huxley was putting together a series of short books on biology. He was keen to have each one written by leading thinkers that would focus on new and emerging principles. Although Elton was just twenty-six, Huxley respected his extensive Arctic experience (three expeditions by that time) and admired the original thinking evident in his published work. Huxley proposed, and Elton accepted to write, a short book on animal ecology.

Elton hurled himself into the project. He wrote like a fiend, most every night from 10 p.m. to 1 a.m. in his flat near Oxford's University Museum and completed the book in just eighty-five days! Despite the breakneck pace, the resulting *Animal Ecology* was a classic in terms of both style and substance. The 200-page book had an engaging, conversational tone, with many handy analogies. Its chapters were logically organized around key ideas that introduced and set the agenda for major facets of what Elton saw as the new science of ecology.

Elton explained that his book was "chiefly concerned with what may be called the sociology and economics of animals." The analogy to human society and economics was deliberate. "It is clear that animals are organized into a complex society, as complex and as fascinating to study as human society," and "subject to economic laws," Elton wrote. The analogy had deep roots in biology. The concept of an "economy of nature" had been embraced by the great naturalist Carl Linnaeus in the eighteenth and by Darwin in the nineteenth century. The connotation was that like human society, animal communities were composed of interacting creatures with different places and roles.

"At first sight we might despair of discovering any general principles regarding animal communities, " Elton said. "But careful study of simple communities," as he had performed in the Arctic, "shows that there are several principles which enable us to analyse an animal community into its parts, and in the light of which much of the apparent complication disappears."

Those principles emerged from the central importance Elton placed on food and food chains. He saw food as the "currency" of animal economies. "The primary driving force of all animals is the necessity of finding the right kind of food and enough of it. Food is the burning question in animal society, and the whole structure and activities of the community are dependent on questions of food-supply," he asserted. Elton encapsulated each principle that flowed from this basic premise with a charming and apt Chinese proverb:

Food Chains
"The large fish eat the small fish; the small fish eat the water insects; the water insects eat plants and mud."

Food chains formed the "economic" connections among the various members of a community. Chains of animals were linked together by food, and all were ultimately dependent on plants. In Elton's scheme, plant-eating herbivores were "the basic class in animal society," the carnivores that preyed upon them were the next class, and the carnivores that preyed upon them the next class and so on, until one reaches an animal "which has no enemies" at the terminus of the food chain.

According to Elton, the structure of these food chains was related to the sizes of creatures at different levels.

Size of Food
"Large fowl cannot eat small grain."

Elton argued that size is the main determinant in the structure of the food chain. Things that were too big to subdue or too small to support an animal were off the menu: "the size of the prey of carnivorous animals is limited in the upward direction by its strength and ability to catch the prey, and in the downward direction by the feasibility of getting enough of the smaller food to satisfy its needs."

These size parameters in turn had important consequences for the numbers of the different kinds of animals in populations and food chains.

The Pyramid of Numbers
"One hill cannot shelter two tigers."

Elton noted that the animals at the base of a food chain are typically abundant, while those at the end, such as tigers, were relatively few in number. There was generally a progressive decrease in numbers between the two extremes. Elton called this pattern the *pyramid of numbers.*

One example he cited was an English oak wood, where one finds "vast numbers of small herbivorous insects like aphids, a large number of spiders and carnivorous ground beetles, a fair number of small warblers, and only one or two hawks." Another example he had documented first-hand was in the Arctic, where copious numbers of crustacea are eaten by fish, the fish eaten by seals, and the seals by a smaller number of polar bears. Elton asserted that such pyramids existed in animal communities "all over the world."

The pyramids of numbers implied that normally there was some balance of animal numbers in a given area. A fundamental question then was how such densities were maintained: How did animals regulate their numbers so that they avoided overpopulation on one hand, and extinction on the other? Elton suggested that, in general, increases in numbers were held in check by predators, pathogens, parasites, and food supply. Extinction was avoided, he explained because as a prey became scarce, predators would switch to other quarry, allowing numbers to recover.

Elton's picture of the regulation of animal numbers, then, was somewhat akin to Cannon's idea of homeostasis—that levels were kept within ranges by counterbalancing factors. (Elton did not use this term, for Cannon had not yet popularized it, but some later ecologists did).

Elton considered the regulation of animal numbers to be of enormous fundamental as well as practical significance, and devoted almost one-quarter of his book to the subject. But, he admitted, "It has been necessary to speak in generalities, since so little is known at present about the rules governing the regulation of animal numbers."

Indeed, ecologists were spurred by Elton's book to search for those rules of the regulation of animal numbers, just as physiologists were motivated by Cannon to search for the rules that regulated processes within humans and other creatures.

And that is where we will head next.

But just before we do, there is one more impact of Elton's book to mention—in fostering the myth of suicidal lemmings. According to Elton's reading of Collett's book, a lemming year occurred "when a whole lot of lemmings apparently went crazy and walked downhill." He wrote in *Animal Ecology*: "The lemmings march chiefly at night, and may traverse more than a hundred miles of country before reaching the sea, into which they plunge unhesitatingly, and continue to swim until they die." This description, however, was based on yarns collected in Collett's book. Elton had never seen a lemming, nor a migration, let alone a mass suicide.

The myth of lemming suicide received a considerable boost from the 1958 Walt Disney film *White Wilderness*, which depicted lemmings leaping to their demise. After the narrator explained, "A kind of compulsion seizes each tiny rodent and, carried along by an unreasoning hysteria," viewers saw lemmings leaping into the water from a high cliff. The scene was faked; the animals were flung off the cliffs by the filmmakers.

The movie won an Academy Award.

PART II
THE LOGIC OF LIFE

Anything that is found to be true of *E. coli*
must also be true of Elephants.

—JACQUES MONOD AND FRANÇOIS JACOB

Elton described how the regulation of animal numbers was enormously important both in nature and in applied fields, and Cannon explained how the regulation of physiology was crucial to animal and human health. By their own admissions, however, neither could say much in detail about how the quantity of anything was regulated in ecosystems or bodies.

The challenges in deciphering the rules of regulation were a bit different for ecologists and physiologists. For Elton and his tribe, the "players" were generally visible to the naked eye—the animals and plants in a given place. But the ecologists lacked ways of getting at the rules of the game; ecology was largely an observational and descriptive enterprise, not an experimental one.

Cannon and his clan, in contrast, were very good at conducting experiments, but they were handicapped because the study of physiology in the 1930s was largely restricted to phenomena observable at the level of body organs and tissues. The players that regulated those phenomena were invisible molecules that were difficult to isolate and identify.

In the next three chapters, I tell stories of how both general and some specific rules of physiological regulation were discovered. Funny enough, the first breakthroughs came from studying creatures without bodies—the tiny bacteria found in our digestive systems (Chapter 3). This pioneering work was important, because, although deciphered in bacteria, the rules turn out to be general ways of regulating all kinds of processes in all sorts of creatures, including ourselves. It was by following the trails blazed by these pioneers that the specific rules for important processes in humans—such as the regulation of cholesterol metabolism (Chapter 4) and cell growth (Chapter 5)—were cracked open. The result of identifying the specific players and rules of those games has been a biomedical revolution that has gone far beyond Cannon's greatest optimism, or imagination.

The discovery of these general rules was important for two additional reasons. First, they underpin what the pioneering molecular biologist François Jacob has described as a "logic" of life. The term is used in both its formal meaning (if A regulates B, and B regulates C, then A regulates C), and in the informal connotation that the regulatory logic makes "sense" for the organism—the same connotation as Cannon's "wisdom of the body." I believe that understanding this logic greatly enhances one's appreciation for how life works.

And the second reason these general rules are important, and one major reason for my writing this book and structuring it as I have, is that analogous rules and logic operate on the ecological scale. I will get to specific ecological rules later in Part Three, but alert you to the importance of and similarity in logic now so that you might pay as much attention to it over the next few chapters as to the particulars of each story.

CHAPTER 3
GENERAL RULES OF REGULATION

The cell thus adapts its work to its wants.
It produces only what it needs when it needs it.

—FRANÇOIS JACOB

Great Britain was not the only nation interested in exploring the poles. Driven by economic and strategic considerations, sometimes by national glory, and occasionally by scientific curiosity, many countries sent expeditions north and south in the first part of the twentieth century.

On July 11, 1934, the three-masted French ship *Pourquoi-Pas? IV* ("Why Not?") left Saint-Malo on the Normandy coast for the icy shores of Greenland. In command was the celebrated polar explorer Jean-Baptiste Charcot. Trained as a physician, Charcot abandoned medicine and made his reputation on two government-sponsored French Antarctic expeditions: on the *Français* in 1903–1905 and the first *Pourquoi-Pas?* in 1908–1910. Braving ice, storms, temperatures that plunged to more than forty degrees below zero, and the long polar night, Charcot discovered new lands, charted over 1,800 miles of coastline and islands, became a national hero, and earned the genuine admiration of fellow explorers. After World War I, he turned his attention to the Arctic. This voyage was the sixty-seven-year-old's twenty-fifth polar expedition and his tenth to Greenland.

FIGURE 3.1 The Pourquoi-Pas? in Greenland, 1934.

Photo by Jacques Monod, © Institut Pasteur/Archives Jacques Monod.

The ship carried a crew of thirty-three men, all volunteers. Also on board were six university students, four of whom were to be dropped off at the village of Angmagssalik to live among the Inuit for a year as part of an ethnographic study. Two others were to conduct scientific studies on board and on shore, including twenty-four-year-old Jacques Monod.

Raised near the famous seaside resort of Cannes, Monod was an experienced sailor but an amateur in comparison to Charcot's team. He had no previous experiences in seas such as those he was about to encounter. The young zoologist had set aside his research at the Sorbonne in Paris for the privilege of joining Charcot's team and the two-month-long adventure of sailing to the Arctic. His duties were similar to Elton and his compatriots—to collect specimens. Twelve days after leaving France, the ship stopped in the fog-enshrouded Faroe Islands. After making some repairs to a damaged boiler, the *Pourquoi-Pas?* sailed on to Iceland, took on coal, and steered for Greenland. All along the way, Monod collected plankton—small crustaceans, marine worms, and larvae—by dragging a net overboard. With each degree of latitude, the air grew colder, and more and more ice began to appear. As the ship approached Scoresby Sound on Greenland's east coast, ice fields blocked her entry. For five days, she inched forward, with a lookout in the crow's nest spotting gaps in the ice that often closed as quickly as they appeared, and another crew member at the rear pushing away blocks of ice that could smash the propeller.

After finally reaching shore, Monod had just three days to collect samples of the coastal marine life before the ship had to leave for Angmagssalik (today Tasiilaq). To collect rock and mineral samples, Monod and a companion set off to climb mountains around the fjord. An avid climber, he was enthralled by all he spied. "I saw so many beautiful and extraordinary things!" he wrote to his parents. "My dears, if you knew in what state of wonder and breathlessness I am!"

With its coal reserves dwindling, the ship had to depart to resupply in Iceland. It was promptly pounded by hurricane force winds. Navigating around icebergs, and with poor visibility, Charcot decided that the ship had to continue at all costs, as its coal supply was nearly exhausted. The crew was able to force their way to Reykjavik, to refuel, and to sail home without further incident.

Monod published a preliminary account of his collections and observations but, alas, he did not become a polar biologist. Two years later he was invited again to join the *Pourquoi-Pas?* on its next voyage to Greenland. Monod was leaning toward going, but at the last moment he decided instead to go to the California Institute of Technology to study genetics in the laboratory of Nobel laureate Thomas Hunt Morgan.

That turned out to be a remarkably fortunate decision. On September 15, 1936, after delivering supplies to Greenland and waiting out a storm, the *Pourquoi-Pas?* stopped once again in Reykjavik before resuming her voyage home. But within hours, she was caught within a fierce storm. Early on September 16, her fore and aft sails were shredded, and the jib smashed the radio antennae. The crippled ship drifted, was pushed onto its side, then smashed on a reef. Charcot and all but one of the forty-four men aboard perished in the cold, roiling seas.

The sparing of Monod's life also turned out to be a fortunate event for biology. Although Monod discovered nothing while in California, he would later become a co-founder of the new field of molecular biology. He and his collaborators would decipher some of the first general rules of the regulation of life at the molecular level, discoveries that would send him north once again—to Stockholm to collect a Nobel Prize.

But he would first have to endure a long and very lethal storm.

GROWTH... INTERRUPTED

After his stay in California, Monod returned to Paris to resume research at the Sorbonne and to search for a problem that he could sink his teeth into that would merit a doctorate.

This was a time for simple questions in biology, as so little was known about the processes going on inside living cells. One of the behaviors characteristic of all cells is the making of more cells by cell division. The questions were very Eltonian: What nutrients did cells require? What determined their numbers?

Before he had left for California, Monod had begun experiments into these matters. But the first organisms he studied were a group of single-celled protozoans that grew very slowly in the laboratory —a poor choice as a research subject. André Lwoff, a microbiologist at the nearby Pasteur Institute, suggested that Monod try bacteria instead, which were easy to grow in culture and multiplied very quickly.

Previously, researchers had largely used ill-defined food for bacterial broths, made, for example, by grinding up cow brains. Monod's first advance was to use carefully defined ingredients that allowed him to undertake a series of experiments in which he systematically varied almost every ingredient and measured the impact on bacterial growth. One of his first clear-cut results was to show that the amount of bacterial growth was directly proportional to the amount of the carbon energy source provided (a sugar, such as glucose or mannitol). This observation implied a very simple relationship between nutrition and growth: bacteria converted whatever food was available into more copies of themselves.

By the summer of 1939, Monod was making progress, but the winds of war were once again blowing across Europe. Like many Frenchmen, Monod did not see what was coming. On August 31, 1939, Monod wrote to his father: "[T]here will be no war. Hitler . . . knows what it would cost him." The very next day, Germany invaded Poland, prompting a declaration of war from France and Great Britain.

War did not break out immediately between France and Germany— days, weeks, and then several tense months passed without combat. Monod became concerned that if fighting did erupt, he would be drafted into some menial administrative work on account of his age (thirty). He wanted to be able to use his scientific talent in some capacity, so he decided to leave the Sorbonne and enlist in the Army to be trained as a communications engineer.

Just after Monod completed his initial training, war finally came when Germany launched a massive assault on the Netherlands, Belgium, and Northern France on May 10, 1940. The French Army was overwhelmed in a matter of days; Monod's regiment never made it off their base until the war was all but over. Monod was again lucky in not being taken as a prisoner of war. After France surrendered, he returned home to resume his research in German-occupied Paris.

WHAT DO BACTERIA LIKE TO EAT?

At thirty years old, Monod was relatively old for a graduate student. He was desperate to find a problem to study that would enable him to finish his doctorate. He had carried out a large number of experiments examining the growth properties of bacteria in broths containing different individual sugars. In the fall of 1940, he decided to

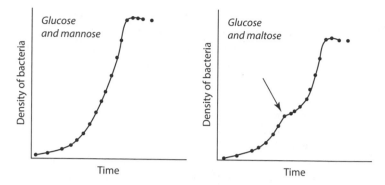

FIGURE 3.2 Monod's double-growth curve. When bacteria were grown in glucose and mannose, a single curve was observed (left), but when bacteria were grown in glucose and maltose, the bacteria grew exponentially, then paused briefly (arrow) before resuming exponential growth (right). That pause and second growth curve became the basis of Monod's thesis and eventual Nobel Prize.

Figure drawn by Leanne Olds based on original data in Jacques Monod's laboratory notebooks.

explore what happened when bacteria were offered different combinations of sugars.

Plotting the concentration of bacteria over time, he saw some familiar-looking growth curves, identical to those he obtained with single sugars. Those curves had three distinct phases: a short lag phase before the growth phase when the bacteria grew exponentially, doubling every thirty to sixty minutes, then a stationary phase when the concentration of bacteria did not increase further. But with certain combinations of sugars, the curves were different. They appeared to have two growth phases, separated by a second lag phase. [Figure 3.2]

Puzzled, Monod showed what he called his "double-growth" curves to Lwoff. The senior scientist hesitated and then said, "That could have something to do with enzyme adaptation."

"Enzyme adaptation? Never heard of it!" Monod replied.

Lwoff gave Monod a few older papers that had reported a phenomenon in which bacteria or yeast cells adapted to the presence of a nutrient by making an enzyme that broke it down. The lags in Monod's curves could be the time needed for the microbes to adapt to each sugar. How a simple microbe "knew" to make an enzyme in response to a specific chemical was a complete mystery—Monod decided on the spot that solving it would be his quest.

Monod discovered that the appearance of a second growth curve depended on which specific sugars were provided. That suggested to him that bacteria preferred to eat certain sugars over others. They were ready to eat some sugars, but needed time to adapt to other, less preferred sugars and to make the enzymes necessary to digest them. He suspected that the explanation for the double-growth curve was that the bacteria were first using up one preferred sugar, then, after a pause, switching to using the second, less preferred sugar.

To test this idea, he had the simple notion of varying the ratios of the amounts of each sugar in the experiment. He reasoned that, if he was correct, the length of the growth phases when each sugar was being consumed would shift accordingly. That was exactly what he saw. [Figure 3.3]

Lwoff was impressed by Monod's gift for designing experiments that zeroed in on each point that he wanted to test. The Sorbonne awarded Monod his degree, although a member of his committee said, "What Monod is doing is of no interest to the Sorbonne."

Monod hoped to study the enzymes the bacteria produced in response to certain sugars, but before he made any headway, his work was interrupted yet again. As the German occupation continued, the atmosphere in Paris grew increasingly tense, and its streets more dangerous for Monod's Jewish wife Odette. She left the capital and took their children to a safer area in the south of France. Anticipating that there would be another battle for France when the Allies reinvaded Europe, Monod decided to join the most militant of Paris' Resistance groups, the Francs-Tireurs et Partisans (FTP).

Monod's responsibilities included gathering intelligence and coordinating arms drops from the Allies. For months, Monod juggled a double life as a Sorbonne scientist and Resistance operative, even hiding incriminating documents in the leg of a stuffed giraffe outside his laboratory. But as German pressure on the Resistance

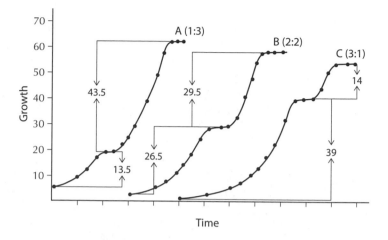

FIGURE 3.3 The proportion of each part of the double-growth curve depends on the ratio of the two sugars. As Monod mixed the two sugars in 1:3, 2:2, and 3:1 ratios, the first growth curve lengthened and the second growth curve shortened proportionately. This revealed that the bacteria used one sugar first, then the second.

From Monod (1942), modified by Leanne Olds.

increased, and some of Monod's superiors and colleagues were arrested and tortured, it became too dangerous for Monod to work at the Sorbonne or to sleep at home. Lwoff offered Monod refuge at the Pasteur, where he was able to continue doing experiments for a few months. Eventually, Monod had to abandon working in the lab. He went completely underground, wore a disguise, and stayed at networks of safe houses. [Figure 3.4]

Monod rose to become a senior officer in the national Resistance organization, the *Forces Françaises de l'Intérieur* (French Forces of the Interior or FFI), where he was involved in coordinating sabotage and even ordering the execution of traitors who collaborated with the enemy. Monod was one of the commanders who helped coordinate the battle for the liberation of Paris in August 1944. He then served as an officer in the French Army until Germany's surrender.

FIGURE 3.4 Jacques Monod's identity card, French Resistance, 1944. Monod held the rank of Commandant in the FFI. His alias "Malivert" is given because Resistance members could not use their real names.

Image courtesy of Oliver Monod.

SEARCHING FOR THE RULES OF ENZYME REGULATION

The war had occupied Monod, his family, and his country for six years. When it finally ended, he was eager to put those dark times behind him and to hurl himself back into research. Lwoff offered, and Monod accepted, a permanent post at the Pasteur Institute.

Monod picked up where he had left off during the war. The logical appeal of enzyme adaptation was irresistible: How did a bacterium, so tiny it was barely visible in a microscope and without any nervous or endocrine system—just a bag of chemicals inside a membrane—"know" to make the right enzyme for whatever sugar was available?

Enzymes are proteins, and cells make thousands of different proteins. Monod recognized that his question was fundamentally a question of regulation: How did a cell "decide" to make one particular enzyme under certain conditions but not others?

Monod believed that there was much more at stake in his research on the regulation of bacterial enzymes than merely questions about the sugar-eating habits of microbes. He understood that what made the different types of cells in more complex creatures distinct from one another was similarly a matter of regulation. For example, red blood cells made hemoglobin proteins that carried oxygen, and white blood cells made antibody proteins that fought infections. Monod believed that understanding why and how a bacterium made a particular enzyme could shed light on the profound general mystery of how different cell types were made.

To try to crack open that mystery, he decided to focus on just one sugar, the milk sugar lactose, and one key player, the bacterial enzyme that cleaves it into galactose and glucose, called "β-galactosidase." Bacteria prefer to use the simple sugar glucose for energy. To use lactose, a compound made of the two sugars glucose and galactose, they have to cleave it into its two halves.

The late 1940s and early 1950s were the dawn of molecular biology, and there was very little precedent for how to do most kinds of experiments. Monod and his team excelled at developing the techniques to sort out different possibilities. The key observation was that the presence of the sugar caused the appearance of the enzyme. One possible explanation for this was that the sugar somehow activated

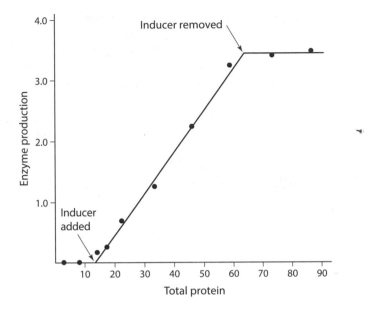

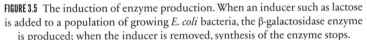

FIGURE 3.5 The induction of enzyme production. When an inducer such as lactose is added to a population of growing *E. coli* bacteria, the β-galactosidase enzyme is produced; when the inducer is removed, synthesis of the enzyme stops.

From Monod and Jacob (1961), redrawn by Leanne Olds.

the enzyme by binding directly to a preexisting, inactive form of the enzyme in the bacteria and converting it into an active form. In a series of clever and technically challenging experiments, Monod and his team ended up blasting this idea apart.

Monod's experiments showed instead that lactose tightly regulated the *production* of the enzyme. When the bacterium *Escherichia coli* was grown in the absence of lactose, there were just a few β-galactosidase enzyme molecules present in a cell. When lactose was added, this rose to several thousand molecules per cell in just a few minutes. When the sugar was removed, the synthesis of the enzyme stopped. [Figure 3.5] This turning on and off of enzyme production was regulated somehow by the presence/absence of the sugar. The sugar was said to be an *inducer* of enzyme production.

This was all very logical on the part of bacteria—it only made the enzyme when lactose (a food source) was present, and it did not waste energy making the enzyme when there was no lactose around. But how did that logic work?

The rules of the regulation of enzyme production would elude Monod for several years. And the main reasons for that were twofold: first, he did not yet know all the players in the game; and second, he had a mental block about how the logic of regulation might operate. The simple observation was that in the presence of the inducing sugar, bacteria made the enzyme. Monod and his collaborators kept thinking of the inducer as something that positively controlled enzyme synthesis, (denoted schematically here and throughout the book with an arrow →):

Inducer (lactose)

↓

Enzyme (β-galactosidase)

To make a breakthrough, they would need to discover another key player and to flip their logic around.

I will explain how they eventually got it right, but the correct logic is so important to understanding regulation and to the entire book that I don't want to risk your getting mired in those experimental details and miss the bigger picture. So I will tell you straight off what Monod was missing and how lactose regulates enzyme synthesis. Then I will back up and tell you how he and his comrades figured it out.

The player that Monod needed to discover was another protein that acted in between lactose and the enzyme. This protein is called a *repressor*, because its job is to specifically repress β-galactosidase enzyme synthesis. The flip in logic comes from realizing that lactose does not positively control enzyme synthesis directly. Rather, lactose inhibits the repressor, so that it no longer represses enzyme production.

Logically speaking, two negatives make a positive.

The double-negative logic of enzyme regulation made great sense biologically: in the absence of lactose, the enzyme that breaks down the sugar is not needed, and a repressor prevents the synthesis of the enzyme (negative regulation denoted by the ⊥ symbol below and hereafter in the book); when lactose is present, it inhibits the repressor, which allows the enzyme gene to turn on and the enzyme breaks down the sugar, providing energy to the cell:

Double-negative logic

No lactose	Lactose
	⊥
Repressor	Repressor (inhibited)
⊥	⊥
Enzyme gene OFF	Enzyme gene ON

Such beautiful logic and economy for just a simple bacterium.

I will get to some of the details of how repression works shortly, but for my purposes here and in the rest of this book, the importance of enzyme regulation is not in the gritty details but in the logic. The breakthrough came from breaking free of a mental bias. When we observe some phenomenon, we are inclined to think of the most direct explanation, with the fewest links in the chain between cause and effect. When we see a car moving down the street, we think somebody is stepping on the gas, not that somebody released the brake.

When the presence of A (e.g., a sugar) leads to the appearance of B (an enzyme), we infer a positive relationship: A causes B. It requires a stretch of the imagination to conjure up the explanation that A inhibits something else (a repressor) that inhibits B.

But it turns out that life—from the molecular scale all the way up to the ecological scale—is usually governed by longer chains of interactions than we first imagine, with more links in between. We need to know about each of those links and the nature of the interactions between them to truly understand, and to intervene in, the rules of regulation on every scale.

To discover the repressor and figure out the logic of enzyme regulation, Monod needed a fresh approach.

DISCOVERING THE REPRESSOR

The fresh approach was to use genetics. Imagine, for instance, that you were interested in how some visible trait was made, say, the pink color of a flower. There are fundamentally two ways you could try to figure out all the players involved in making that pink color. You could take the biochemical approach, which would be to grind up the flower and try to purify all the enzymes that work in making a pink pigment from some simpler chemicals. That turns out to be very difficult and time consuming.

Or, you could take a genetic approach. That would entail taking seeds from some pink plants; and growing thousands of seedlings; and looking for the rare ones that could not make pink flowers but, say, only white ones. Every white plant has some genetic defect, a mutation, in some gene that is involved in the making of the pink pigment. You would then study those genes.

The great advantages to the genetic approach are that it uses a simple visual test to find mutations in genes of interest and that it is unbiased—it makes no assumptions about the number of players or what they do. It can discover players that are not enzymes, for example. Many of the key breakthroughs in biology and medicine over the past half-century were catalyzed by a genetic approach (I will describe two medically important examples in the next two chapters).

Monod and his team went looking for mutations in bacteria that disrupted β-galactosidase production. They isolated two types of mutants. One type made a defective β-galactosidase enzyme: those were mutations in the gene encoding the enzyme itself. This type was expected. But a second type of mutant was especially interesting: these mutant bacteria did not require any lactose to make the enzyme. They made the enzyme all the time ("constitutively"), whether lactose was around or not. In this mutant, the normal on/off regulation of the enzyme was broken. The constitutive mutations were in a separate gene from the enzyme and somehow disrupted the regulation of the enzyme gene.

Understanding how this new player worked would be key to understanding the regulation of the enzyme. But Monod was stumped at first. He had interpreted the constitutive mutants through the logic of the inducer acting as a positive regulator of enzyme production. He reasoned that, if the mutant bacteria required no added inducer to produce the enzyme, then the mutants must make their own internal inducer of β-galactosidase. It would take a new partner to reveal that Monod's logic was faulty.

DISCOVERING DOUBLE-NEGATIVE LOGIC

That new partner was François Jacob. Originally planning on becoming a surgeon before the war, Jacob's career was derailed when he was severely wounded in Normandy while serving as a medic. He went into scientific research instead, and wound up by chance in Lwoff's lab just down the hall from Monod. He was studying a different phenomenon in which bacterial viruses hide out quietly inside bacterial cells until something triggers them to multiply and burst forth. In a short period of time, Jacob had developed important techniques for studying genes in bacteria. He teamed up with Monod in 1957, and one new method from his bag of genetic tricks finally cracked the logic of enzyme regulation.

Unlike humans and most animals, which have two copies of each chromosome (one from each parent) and two copies of most genes, *E. coli* has a single chromosome with one copy of each gene. One trick that Jacob had pioneered was a way to transfer genes between bacteria. This allowed him to construct bacteria that had extra copies of genes and to test how bacteria behaved when mutant and normal genes were mixed together. If Monod was right about the constitutive mutants, then when a normal copy and a mutant copy of the gene were put together in the same bacterial cell, the prediction was that the internal inducer would be produced and the enzyme would be made constitutively.

But when Jacob and visiting American scientist Arthur Pardee did the experiment, they got exactly the opposite result: the bacteria required the inducer (lactose) to make the enzyme. The researchers were stumped at first. Perhaps they had made some technical mistake? But that was not the case: the same result was obtained when they repeated the experiment.

If their technique was not faulty, than perhaps their logic was. Indeed, that is exactly what Leo Szilard, a physicist-turned-biologist and frequent visitor to the Pasteur, suggested to Monod and Jacob. Maybe they were thinking about the inducer in the wrong way? Maybe the inducer did not activate enzyme synthesis directly, as Monod thought, but rather it *inhibited* a *negative regulator* of enzyme synthesis?

Bingo. The double-negative logic made sense of all of their results. The constitutive mutants were not making an internal inducer, they were mutants that lacked a player in enzyme regulation—a *repressor* of enzyme synthesis. The absence of the repressor in the mutant allowed enzyme synthesis to take place continuously without an inducer. And when a bacterium had one good copy of the repressor gene and one mutant copy of the repressor gene, the good copy prevailed and repressed enzyme production, unless an inducer was added.

Once Monod and Jacob overcame that bias of simple positive cause and effect relationships, they started thinking in new ways and seeing connections that had not and would not have occurred to them (or anyone else) before.

One Sunday afternoon, while sitting in a Paris movie theatre with his wife, Jacob's mind started to drift from the film to the puzzle he had been working on for years. Some bacteria harbored hidden viruses that could be activated when bombarded with ultraviolet light. How that worked had stumped Jacob, and no one thought there was any connection between what he was studying and what Monod was researching at the other end of the hallway. Until, in the darkened cinema, Jacob began to picture the virus with its many genes somehow being kept off inside the bacterium.

Then he had a flash of insight—the logic of virus activation was the same double-negative logic of enzyme induction. A repressor must also be keeping the viral genes repressed, until ultraviolet light destroyed or removed it, and the virus genes turned on.

What appeared to be positive activation was again the inhibition of repression.

Virus silent	*Virus activated*
	UV light
	$\perp$
Repressor	Repressor
$\perp$	$\perp$
Viral genes OFF	Viral genes ON

Convinced by what were once believed to be two entirely different phenomena, Monod and Jacob proposed that there were fundamentally two kinds of proteins inside cells; *structural* proteins, such as enzymes that carry out the chemical reactions in cells or build the parts of a virus; and *regulatory* proteins that controlled which structural proteins were made or were not made, depending on conditions. When it came to regulation then, not all proteins were equal. Some proteins were dedicated to controlling others. Monod and Jacob began seeing negative regulation everywhere and finding it at work in other ways.

FEEDBACK

In addition to breaking down nutrients into useful compounds, bacteria and other organisms are also able to build important compounds out of simpler ingredients. The proteins that do all the work in living things are constructed from building blocks called amino acids. When bacteria are grown in a basic medium containing glucose and carbon dioxide as carbon sources, they can make all twenty kinds of amino acids.

However, when specific amino acids are provided to the bacteria, the biosynthesis of that particular amino acid stops quickly. That rapid response suggests that when an amino acid is plentiful, bacteria have some mechanism for specifically shutting off the enzymes that synthesize it.

In the 1950s, many biochemists were busy deciphering the ways that various amino acids were manufactured. They were finding that the synthesis of every amino acid usually involves a several-step-long "pathway" in which an initial chemical precursor (P) is modified by a series of enzymatic reactions into the amino acid. These pathways are drawn schematically as a chain of intermediate reaction products (I1, I2, etc.), each produced by a different enzyme:

$$P \rightarrow I1 \rightarrow I2 \ldots \rightarrow \text{amino acid}$$

It was discovered, for example, that when the amino acid tryptophan was provided to bacteria, the synthesis of an intermediate was halted. This suggested that tryptophan acted on an early-acting enzyme in the pathway. Similarly, it was found that providing the amino acid isoleucine also inhibited the activity of the first enzyme in its synthetic pathway.

These discoveries inspired the concept of *negative feedback*, whereby compounds feed back on their own synthesis as a way of controlling their levels in cells. The study of all sorts of biosynthetic pathways subsequently revealed that not only was negative feedback widespread, but that it also almost invariably operated by the end-product of the pathway directly inhibiting the first enzyme in its pathway.

Like the double-negative logic of enzyme induction, the logic of negative feedback in biosynthetic pathways also made great biological sense: when the end-product of a pathway is abundant, cells do not waste energy making it or any intermediates; but when the concentration is low, the synthetic machinery is not inhibited, and the needed product is synthesized.

These pioneering studies of bacteria revealed four basic ways that one molecule can affect the abundance of another molecule. They constitute a set of general rules and a logic of regulation that, as we shall see, govern all sorts of processes in other species. (You may want to bookmark this page.)

GENERAL RULES OF REGULATION AND THE LOGIC OF LIFE

Positive regulation

A → B   A positively regulates the abundance or activity of B

Negative regulation

A ⊣ B   A negatively regulates the abundance or activity of B

Double-negative logic

A ⊣ B ⊣ C   A negatively regulates B, which negatively regulates C; A increases the abundance of C through double-negative logic

Feedback regulation

A → B → C   The accumulation of C feeds back to negatively regulate A and the production of B and C

THE SECOND SECRET OF LIFE

The discovery of repressors and feedback inhibition prompted intense interest in understanding exactly how these two kinds of regulation worked at the molecular level. What did a repressor do? How did inducers work? How did feedback occur?

Late one evening in the fall of 1961, Jacques Monod walked into the laboratory of his colleague Agnes Ullmann. Usually well-dressed and energetic, Monod's tie was loose, and he looked tired and worried. After a long silence, he told Ullman, "I think I have discovered the second secret of life."

Ullmann thought Monod did not look well, so she suggested that he sit down and have a scotch, their favorite drink. After another drink or two, Monod stood back up and started a long explanation. He was not ill. He was in top form. He recapped years of observations about repression and feedback inhibition, then offered a single unifying explanation for *both* phenomena.

Monod's breakthrough came from picturing the shapes and sizes of molecules. He was thinking about an enzyme his lab was then studying. Enzymes are large proteins, more than one hundred times larger than the substances they work on (called *substrates*), such as sugars or amino acids. Like a key fitting into a lock, substrates fit snugly into a cavity in the enzyme called the active site, where they get cleaved or modified.

The enzyme Monod was studying is the first enzyme in a pathway that makes the amino acid isoleucine. It works on a substrate called threonine and is inhibited by isoleucine, the final product of the pathway. Monod was trying to picture how the small isoleucine molecule might fit into the active pocket in the enzyme and stop it from working. But then it struck him, isoleucine is not the same shape as threonine. Maybe it can't fit into the cavity?

Then he thought about other feedback-inhibited enzymes and realized the same was true for them: they were inhibited by molecules that looked very different from their substrates. What could that mean? Monod figured the place where the feedback inhibitor bound must be different from the active site. The enzyme—the "lock"—must have two keyholes: one for the substrate and one for the inhibitor.

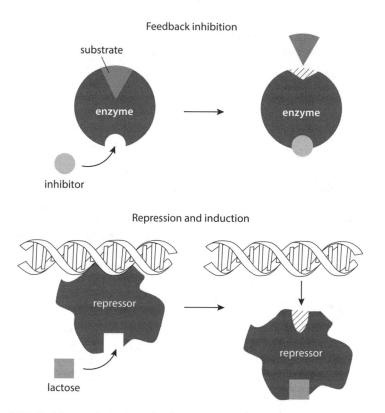

FIGURE 3.6 Allostery is the basis for feedback inhibition and enzyme induction. (Top) The substrate fits into the active site of the enzyme; the inhibitor into a different pocket. When the inhibitor is bound, the shape of the active site changes such that the substrate no longer fits. (Bottom) One site on the repressor binds DNA, another site binds lactose. When lactose is bound, the shape of the repressor changes, and it no longer binds DNA, which allows the enzyme gene to turn on.

Illustration by Leanne Olds.

Somehow, the binding of the inhibitor altered the shape of the enzyme in a way that it could no longer bind its substrate (that keyhole was closed). Monod dubbed this phenomenon *allostery* (from the Greek *allos*, meaning "other," and *stereos*, meaning "solid or object.") He thought allostery could be an important way to regulate the activity of proteins (see Figure 3.6, top).

Then, that one evening, all the pieces fell into place. The inducer and repressor worked in exactly the same way as feedback inhibition, by allostery. The repressor must also have two sites: one site for binding DNA, one for the inducer. When no inducer is present, the repressor binds to DNA, keeping the gene off; when the inducer is present and binds to the repressor, it would induce a change in the physical shape of the repressor, causing it to fall off the DNA and allowing the gene to turn on (see Figure 3.6, bottom).

Monod had two lines of evidence for one simple but big unifying idea: small molecules (amino acids, inducers) regulate the shape and activity of big molecules (proteins). Having connected the seemingly unrelated phenomena of enzyme repression and feedback inhibition, Monod imagined the potential generalities. Allostery could explain, for example, how small molecules like hormones and neurotransmitters regulate the endocrine and nervous systems. Staggered by the potential scope of his idea, he wandered in to test it on Ullmann.

Since DNA was the first secret of life, then perhaps allostery—with all its implications for understanding how genes and proteins were regulated—was the second. At the very least, the Nobel Committee thought that it and all of Monod's and Jacob's discoveries merited the 1965 Nobel Prize in Physiology or Medicine.

E. COLI AND ELEPHANTS

The importance of Monod's and Jacob's research was not a matter of the specifics of solving the mystery of β-galactosidase enzyme regulation in E. coli. Like Elton and Cannon, the power of their ideas stemmed from their originality and generality concerning the rules of regulation they uncovered.

Just as Elton conceived of ecosystems as a society of organisms interacting with one another through food chains, and Cannon saw the body composed of a collection of organs communicating with one another through the nervous and endocrine systems, Monod and Jacob pictured life in cells as a "society of macromolecules bound together by a complex system of communications regulating both their synthesis and activity."

Monod and Jacob eloquently explained how their insights, derived entirely from the study of single-celled bacteria, had implications for understanding complex phenomena in much more complex organisms. In a masterful synthesis of the state of knowledge in 1961, they quipped that it was a "well-known axiom that anything found to be true of *E. coli* must be true of Elephants."

That was more of a bold wish than a proven or accepted fact, but that did not restrain their speculations. While admitting that regulation in higher organisms might be "immeasurably" more complex, they suggested:

> On the other hand, it seems very unlikely that the main mechanisms recognized in lower forms: allosteric inhibition, induction and repression, should not also be used in differentiated organisms. But it is clear that these mechanisms, by their very nature, can be adapted to widely different situations, and would serve entirely different purposes in *E. coli* and Man, respectively.

Not only these mechanisms, but the logic of negative regulation also appeared to Monod and Jacob to be of utmost importance in higher organisms. Recognizing that cancer cells have lost their sensitivity "to the conditions which control multiplication in normal tissues," they suggested that cancers may result from genetic mutations or other agents that inactivate a repressor involved in the control of cell multiplication.

As I will show in Chapters 4 and 5, their speculations turned out to be highly influential and remarkably prescient.

CHAPTER 4

FAT, FEEDBACK, AND A MIRACLE FUNGUS

Rather than replacing genes, we can exploit regulatory
principles to make good genes work harder.

—DR. JOSEPH GOLDSTEIN AND DR. MICHAEL BROWN, TO DR. ROY VAGELOS, CEO OF MERCK & CO.

On June 29, 1935, American Ancel Keys and Englishman Bryan Matthews made camp near the summit of Mount Aucanquilcha, over 20,000 feet above sea level in northern Chile. They built a simple snow shelter by putting up a few poles and draping blankets over them, then crawled inside to escape the wind and the temperatures that plunged to fifty degrees below zero overnight. They remained above 20,000 feet for fifteen straight days, during which they ascended the summit several times. At the time, their achievement was one of the highest conquests of the Andes. But these intrepid mountaineers were not professional climbers—they were academic physiologists.

Keys was from Harvard, and Matthews from Cambridge University. The two were part of the ten-member International High Altitude Expedition (IHAE) that had traveled to Chile to study how the human body adjusted to very high altitude. Aucanquilcha was home to the world's highest permanent population at 17,500 feet, and the world's highest mine at 19,000 feet. The expedition was the largest,

FIGURE 4.1 Ancel Keys in the Andes. Keys (lying down on his back) is having his blood drawn at 20,140 feet to measure how the body responds to high altitude.

From Keys, A. (1936) "The Physiology of Life at High Altitudes." *Scientific Monthly* 43(4): 309.

highest, longest, scientifically best-equipped, and technically most sophisticated effort to understand how humans could live and work under such extreme conditions.

If one sign of a great scientist is the courage to go wherever one's curiosity leads—Cannon, Elton, and Monod are good examples— then Keys might well be the poster child for the entire tribe. A gifted child growing up in California, Keys left high school at fifteen to shovel bat manure in a cave in Arizona, then worked in a Colorado gold mine as a "powder monkey" delivering explosives to miners. After returning and finishing high school, he started out studying chemistry at college, became disenchanted, and left to work as an oiler on an ocean liner that sailed to China. After surviving on a diet that was "mainly alcohol," Keys returned to college, earned a degree in economics, and then another in biology in just six months. He then went to the Scripps Institute in La Jolla, California, where he earned a PhD in oceanography and biology, and then to Cambridge University, where he received another PhD in physiology, before joining the Harvard "Fatigue Laboratory" and organizing the IHAE adventure in Chile.

A second common ingredient in a scientist's journey is serendipity—Cannon's angry cats, Elton's lemming book in that Tromsø shop, and Monod's odd growth curve. For Keys, it was a call from the Army. For six days during his stay below Aucanquilcha's summit, he and Matthews had to get by on nothing more than water and condensed food. Apparently, that was sufficient experience to attract the interest of the quartermaster of the Army. As war broke out in Europe, the Army thought it wise to develop some sort of lightweight, nonperishable food ration, something paratroopers might live on before ground troops caught up to them. So, the Quartermaster Corps of the Defense Department called on Keys for advice.

Keys had moved once more—to the University of Minnesota. A colonel came to Minneapolis, and he and Keys went shopping at the best grocery store in the Twin Cities. They then divided the food up into paper bags and took them to a local military base for testing. After further trials at Fort Benning, Georgia, the ingredients of the inaugural, roughly 3,000-calorie package were decided, including: a piece of hard sausage or canned meat, dry biscuits, a block of chocolate, a stick of chewing gum, matches, and a couple of cigarettes, all in a waterproof package that could fit in a uniform pocket. Dubbed the K-ration (purportedly for "Keys"), over 100 million were produced at the height of the war in 1944.

After the war, Keys turned his attention to other challenges. He was intrigued by statistics from food-starved Europe that showed a dramatic decrease in deaths from heart disease, whereas many prominent American men were dropping from heart attacks. Why did some men get heart attacks and others didn't? Keys recruited 281 Minneapolis-area men aged forty-four to fifty-five into a long-term study of how sixty different characteristics, including diet, affected their risk for heart attacks.

While the Minnesota study was under way, Keys traveled the world, talking about heart disease. When a colleague from Naples claimed that it was not a major problem there, Keys was skeptical, and went to investigate. Studying a group of Neapolitan firemen, he found much lower levels of cholesterol in their blood than in

American businessmen. He discovered the same was true of poor people in Spain. To Keys the correlation was obvious—richer people were eating diets rich in fats and having more heart attacks. But medical colleagues were skeptical of the links between diet, serum cholesterol, and heart attacks. So, Keys and collaborators organized an unprecedented, large-scale international study of heart disease risks in more than 12,000 men from various parts of the world—Yugoslavia, Italy, Greece, Finland, the Netherlands, Japan, and the United States—with very different diets. The "Seven Countries" study was launched in 1958, and the men were to be examined every five years.

In 1963, results were obtained from both studies. After following the Minnesota businessmen for fifteen years, Keys identified one major risk factor for heart disease: serum cholesterol levels. Men with levels greater than 260 milligrams of cholesterol per 100 milliliters of blood had five times the heart attack risk of those with levels below 200. The Seven Countries study found the same at the five-year mark. For example, the average cholesterol level of east Finlanders was 270, and they had more than four times as many heart attack deaths as Croatians with cholesterol levels below 200.

Keys now had strong evidence that what people were eating was making them sick. It had been known for fifty years that atherosclerotic plaques from human aorta tissue contained at least twenty times as much cholesterol as normal aortas, and that feeding cholesterol to animals could induce hypercholesterolemia and atherosclerosis. But it was these large-scale epidemiological studies that helped cement the link between serum cholesterol levels and heart disease in humans and raise awareness of the risk.

What these correlations did not explain is how to treat people with heart disease so that their health might be improved. Cholesterol is not simply some threat to be eliminated: the cholesterol molecule is vital to life. It is an essential component of the membranes of all animal cells that helps maintain the barrier between cells and their environment. Cholesterol modulates the fluidity of the membrane and the mobility of other molecules within it. Moreover, cholesterol is a member of an important class of molecules called sterols and is the precursor of five kinds of steroid hormones, including cortisol, the sex hormones testosterone and estrogen, and a component of bile critical for digestion. So, the important issue is how to maintain healthy levels of the sterol—cholesterol homeostasis. In the early 1960s, heart disease was by far the leading cause of death in American adults. Changing that situation, if possible, would require knowledge of the rules of cholesterol regulation.

The most critical insights into cholesterol regulation came when two young physicians, influenced by the ideas of Monod and Jacob, took several pages right out of the Frenchmen's playbook. First, they teamed up to tackle the problem. Second, they decided to study people in whom, like the constitutive bacterial mutants Monod and Jacob studied, the regulation of enzyme synthesis was broken. Then, by analyzing these human mutations, they methodically worked out the logic of cholesterol regulation. And, exactly twenty years after Monod and Jacob, they made that same trip to Stockholm to collect their Nobel Prizes.

DISCOVERING FEEDBACK

Joe Goldstein and Mike Brown first met in 1966 as interns at Massachusetts General Hospital in Boston, when they began their rotations working in the emergency room. Although the two came from different backgrounds—Goldstein grew up in a small town in South Carolina, Brown in New York and Philadelphia—the two hit it off right away. After long days on the wards, they found that, more so than the other young physicians, they each liked to sit around and discuss the possible pathologies of the diseases they had seen.

After their stints in Boston, they both moved to the National Institutes of Health (NIH) in Bethesda, Maryland, as clinical associates, where they had double duties, doing basic research and seeing patients. Goldstein's clinical assignment was in the National Heart Institute. Two of the first patients he saw were extraordinary—a six-year-old girl and her eight-year-old brother who had both suffered heart attacks. It was a life-changing moment for Goldstein.

The siblings had come to the NIH hospital because they had a condition known as familial hypercholesterolemia (FH). The inherited disease occurs in two forms: in the *heterozygous* form, found in about one out of 500 people, individuals have one copy of the mutant gene, serum cholesterol levels in the 300–400 range, and suffer heart attacks as early as age thirty-five; in the very rare *homozygous* form (about one in 1 million people), individuals have two copies of the mutant gene, have astronomical serum cholesterol levels in the 800 range, and suffer heart attacks beginning as early as five years old.

The Texas siblings had the most severe, homozygous form. Goldstein told Brown about the children, and they started wondering what kind of defect could cause such a dramatic spike in cholesterol levels. Parts of their very full days at the NIH were spent attending night courses on various subjects, including one in which Monod's and Jacob's new ideas about regulation were discussed at length. Goldstein and Brown had learned in medical school that cholesterol synthesis was subject to feedback regulation: when dogs were fed a high-cholesterol diet, cholesterol synthesis was suppressed. Perhaps, Goldstein and Brown wondered, the FH patients had a defect in the feedback regulation of cholesterol?

While most of their talented peers were planning research careers in cancer, neuroscience, and other prestigious fields, Goldstein and Brown decided that they would team up and focus on cholesterol regulation. "It's just an amorphous glob," their friends teased. But Goldstein and Brown ignored those jabs, and after moving to the University of Texas Southwestern Medical Center, they joined forces officially by merging their laboratories. In just two years, working seven days a week, they cracked open the mysteries of cholesterolemia and the logic of cholesterol regulation in a series of elegant experiments.

At the time they began their work, the pathway through which the twenty-seven-carbon cholesterol molecule is built up from a two-carbon precursor had been deciphered—an achievement that had earned a total of eleven Nobel Prizes for a series of discoverers. The pathway involved about thirty enzymes, but it was known that the rate of cholesterol synthesis was determined by the activity of the enzyme that acted at the first step of the pathway, whose name is a mouthful: 3-hydroxy-3-methylglutaryl coenzyme A reductase. I'll just call it "the reductase," as it is the one and only enzyme I am going to talk about in this chapter, and what it does exactly is not essential to know here. The important stuff is again the logic of its regulation.

Brown and Goldstein needed to study the activity of the reductase in humans, but because the enzyme is active in the liver, that would be difficult to do in people. Instead, they developed a way of monitoring the enzyme in cells that were taken from people and cultured in the lab. To grow in culture, cells need nutrients that are usually provided in the form of serum. One of the first things that Brown and Goldstein discovered is that the activity of the reductase was negatively regulated by something in serum: when serum was present, activity went down; when serum was removed, the activity went up tenfold.

They then wanted to find out what component of the serum suppressed reductase activity. They suspected that it was some lipid-containing component of the serum, so they tested the activity of the so-called LDL (low-density lipoprotein), HDL (high-density lipoprotein), and non-lipid-containing fractions. They discovered that LDL, but not HDL or other fractions, was a potent inhibitor of enzyme activity:

Following the logic and discoveries of Monod and Jacob, Goldstein and Brown hypothesized that hypercholesterolemia patients who overproduce cholesterol might have mutations in the reductase gene that made the enzyme resistant to regulation by LDL. Their first set of measurements seemed to support that case. When cells from FH patients were grown in culture, Goldstein and Brown observed that they had forty- to sixtyfold greater reductase activity than cells from healthy people, and that LDL had no effect on their enzyme activity:

Normal controls	*FH patients*
LDL	LDL
⊥	⊥
Reductase	**Reductase (↑~sixtyfold)**

But their next experiment nixed the idea that the reductase itself was altered in FH patients and raised a different possibility. LDL is a particle consisting of both protein and lipids, including cholesterol. Brown and Goldstein hypothesized that cholesterol was the active agent in suppressing enzyme activity, so they fed cells cholesterol that was free of lipoprotein. They found that cholesterol was indeed a potent inhibitor of enzyme activity in normal cells, but they were surprised to discover that cholesterol also inhibited reductase activity in FH cells. This revealed that the FH patients' reductase was just as sensitive to feedback control by cholesterol as healthy people's, but not when the cholesterol was in the form of the LDL particle.

Since the defect in FH patients was not in their reductase enzyme, it had to be in something else, some player that Brown and Goldstein did not know about. Because the cholesterol-carrying LDL circulates outside cells, perhaps the transfer of cholesterol from LDL outside cells to inside them was defective in FH patients?

Brown and Goldstein imagined that there might be a specific receptor for LDL on the outside of cells. To test this idea, they did a simple experiment. They attached a radioactive "label" to LDL particles, so that they could measure their binding to cells. They observed that labeled LDL particles were bound strongly by normal cells but not by FH cells. This experiment revealed a specific receptor for LDL on normal cells, but this receptor was missing from FH patient's cells. There was indeed another player in the regulation of cholesterol levels.

Brown and Goldstein figured out how the receptor worked to bring cholesterol from outside to the inside of cells. The protein part of LDL carried the cholesterol and docked with receptor, the cholesterol molecules were then separated from the protein inside cells, where they could act to regulate reductase activity. The discovery of the LDL receptor explained why circulating LDL could not regulate cholesterol synthesis in FH patients as it did in normal people:

Normal controls	FH patients
LDL (cholesterol)	LDL (cholesterol)
|	|
LDL receptor	LDL receptor
⊥	⊥
Reductase	**Reductase (↑ ~sixtyfold)**

Brown and Goldstein also discovered that the number of LDL receptors on cells is feedback regulated just like the reductase: when cholesterol levels are low in cells, the number of LDL receptors and reductase activity is increased; when cholesterol levels are high, the number of receptors and reductase activity is decreased. The logic makes perfect sense in enabling cells to maintain their levels of cholesterol: when cholesterol levels are low, the cells will draw cholesterol out of the bloodstream via the LDL receptor as well as synthesize it; when levels are sufficient, both the reductase and the LDL receptor are suppressed.

More than 93 percent of all of the body's cholesterol is contained in cells, where it serves vital functions. But 7 percent is in circulation, about two-thirds of it in the form of LDL, about one-third in the form of HDL. Epidemiological studies and animal experiments had shown that it is the circulating LDL form of cholesterol ("bad cholesterol") that is the major culprit in plaque formation and heart disease. Could Brown's and Goldstein's insights into the rules of cholesterol regulation be brought to bear in treating disease? Unbeknownst to the duo, the seeds of a medical revolution were already being planted far away from Texas.

A "PENICILLIN" FOR CHOLESTEROL?

Akira Endo grew up on a farm in Akita, Japan, with his extended family. His grandfather shared his interest in medicine and science by teaching young Akira about nature. By age ten, Endo was fascinated with mushrooms and molds. He learned, for example, about a mushroom that killed flies, but not people. In college, he learned about the pioneering work of Alexander Fleming, who discovered the antibiotic penicillin produced by the blue-green fungus *Penicillium*.

After graduation, Endo joined the Sankyo pharmaceutical company in Tokyo, where he first worked on food ingredients. Charged with the task of finding an enzyme that could reduce the pulp in wines and ciders, Endo searched through more than 200 strains of fungi. He identified a parasitic fungus that grew on grapes and made the right enzymes. After the product was successfully commercialized, Endo turned to a new interest—cholesterol.

As the epidemiological studies linking high cholesterol to heart disease became widely reported, Endo, like scientists at many pharmaceutical companies, thought that inhibitors of cholesterol synthesis could be very important drugs. And indeed, many drugs were developed in the 1960s aimed at combating high cholesterol, but they were largely ineffective and typically plagued by side effects. None had targeted the reductase.

But Endo had an original idea and a different approach. He knew very well that fungi produced all sorts of compounds, such as penicillin, that thwarted the growth of competing microbes. He also knew that in some fungi, the molecule ergosterol was the major sterol in membranes, not cholesterol. So, he reasoned, perhaps some fungus had naturally evolved a compound that could inhibit cholesterol synthesis in other organisms. Could he find such a "penicillin" for cholesterol synthesis?

To undertake the search, Endo devised a simple strategy. He knew that the reductase was the first step in the cholesterol pathway, so he designed a straightforward test to detect anything that inhibited enzyme activity. He could then sample the broths of fungi grown in the laboratory to see whether any made inhibitors of the reductase. In April 1971, he and three assistants started their quest.

Day in and day out, Endo and his team grew and tested *a lot* of fungi, about 6,000 in all. After two years of searching, they came up with two promising species. They then worked to isolate the active ingredients in their broths. The first, from the fungus *Pythium ultimum*, turned out to be a previously identified antibiotic called citrinin. It inhibited the reductase, but it was already known to be highly toxic to animals. The second, isolated in the summer of 1973, came from the fungus *Penicillium citrinum*, a blue-green mold that had been isolated from the rice of a Kyoto street vendor and a relative of the fungus from which penicillin had been isolated.

To obtain enough of the active ingredient for laboratory studies, they grew a giant 600-liter batch of the fungus, from which they purified a grand total of twenty-three milligrams of the compound, much less than a typical tablet of aspirin. With that precious material, they were able to show that the molecule they called ML-236B (later called compactin) was a potent inhibitor of the reductase, effective at low concentrations. Part of the compactin molecule closely resembled the normal substrate of the reductase, which explained how it inhibited the enzyme: it slipped into the active site ("lock") of the reductase where the substrate ("key") normally fit and blocked the enzyme from working.

Compactin looked like a promising drug. And now, when I tell you that what Endo had discovered was the first statin, a class of drugs taken today by over 25 million people with global sales of $29 billion in 2012, you would think that Endo might have become rich and famous, maybe even won a Nobel Prize?

Nope. Nothing like that happened.

The road from the discovery of compactin to a medicine for heart disease would turn out to be twisting and bumpy. The story is a tale about the convictions and perseverance of several people, including Endo, Brown, Goldstein, and a few pharmaceutical executives—although not those at Sankyo.

FROM FUNGUS TO THE PHARMACY

Endo and Sankyo published their discovery of compactin and patented their new compound. The next crucial step was to test compactin's activity in animals. The good news was that it did not cause any apparent harm to rats. The bad news was that feeding it to rats for seven days caused no changes in cholesterol, nor did the administration of high doses of the compound for five weeks. Compactin was also ineffective in mice. With these negative results in animals, it appeared that compactin's prospects as a drug and Endo's years of hard work had hit a dead end.

Endo, however, did not give up on compactin. One night in the spring of 1976, he was at a bar near work when he ran into a colleague who was using laying hens in his research. Some further experiments Endo had done gave him reason to hope that the negative results in rats and mice had to do with the peculiarities of how cholesterol was regulated in those species, and that the drug might be effective in other animals. His colleague agreed to let Endo test the drug in his birds.

In just one month of treatment, the birds' cholesterol levels dropped by 50 percent! Endo's success in hens prompted tests in monkeys and dogs, where compactin caused 30–44 percent reductions in cholesterol. Given the close biological relationship between monkeys and humans, these results boded well for the drug's potential. Sankyo formed a Compactin Development Project team composed of

Endo, pharmacologists, pathologists, chemists, and toxicologists to further develop the drug.

But just as compactin's future had begun to look brighter, the toxicologists saw some abnormalities in the liver cells of rats fed very large doses of the drug. It took many months before it was decided to continue with clinical development. However, once testing in humans was under way, Sankyo's toxicologists found another problem. Dogs that had been fed large doses of compactin for two years developed what they believed to be intestinal tumors. Sankyo terminated the development of compactin in August 1980.

By this time, others had become aware of Endo's and Sankyo's work. Roy Vagelos, an accomplished lipid biologist, was the head of research of the Merck pharmaceutical company in the United States. Vagelos had gone to Merck from academia hoping to change the way that drugs were discovered. For decades, companies had searched for drug candidates by screening large numbers of compounds for activities against cells or microbes, rather than against a particular molecular target. Vagelos aimed to use biochemistry to devise more targeted approaches. He also happened to have spent an exciting year working with Jacques Monod in Paris, so he was attuned to thinking about the logic of regulation. Between Brown's and Goldstein's work on the rules of cholesterol regulation and Endo's discovery of a natural reductase inhibitor in fungi, Vagelos recognized the potential for a new kind of cholesterol drug.

Vagelos encouraged Merck scientists to search for compactin-like substances in other fungi, and by early 1979, they had discovered a similar compound in the fungus *Aspergillus terreus*. Later named "lovastatin," it differed from compactin only by the presence of an additional methyl group (one carbon and three hydrogen atoms) on the molecule. Merck promptly launched a clinical trial of lovastatin in humans, but when Vagelos heard about Sankyo's decision to stop their trial of compactin and rumors of tumors in dogs, he immediately halted Merck's study as well. Compactin and lovastatin were dead in the water, and would have remained so were it not for an unexpected result in Texas.

Brown and Goldstein had also learned of Endo's discovery. Impressed by the potency of the enzyme inhibitor, they asked for a sample of the compound. They also invited Endo to visit their lab in Texas and collaborated with him on a study of compactin's effects. The scientists were quite surprised to observe that compactin not only inhibited the activity of the reductase, but also that cells made a lot more of the enzyme when treated with the drug. This revealed some important double-negative logic in the rules of cholesterol regulation: when internal cholesterol synthesis was inhibited, so was the feedback repression of enzyme synthesis:

Compactin
⊥
inhibits reductase activity → → → cholesterol
⊥ *less* feedback repression
more **reductase**

And now knowing these specific rules, an exciting possibility occurred to Brown and Goldstein. Since they had previously discovered that the reductase and LDL receptors on cells were regulated in unison, they reasoned that compounds that inhibit the reductase enzyme might also *increase* LDL receptor levels. And if that were true, then the increased numbers of LDL receptors on cells might pull more LDL out of the bloodstream, thereby *lowering* circulating LDL cholesterol levels—the crucial factor in heart disease:

	Without compactin	*With compactin*
low	LDL receptor	↑ **LDL receptor**
high	serum LDL cholesterol	↓ Serum LDL cholesterol

To test this possibility, Brown and Goldstein obtained a small amount of lovastatin from Merck and gave it to dogs. Sure enough, the drug increased both the LDL receptor levels and the clearance of LDL from the bloodstream. The results convinced Brown and Goldstein that the drug might be able to clear and lower LDL levels in humans, but by then both Sankyo and Merck had stopped their clinical testing because of the worries about tumors in dogs.

Goldstein went to Japan and visited Endo, who by that time had left Sankyo and joined Tokyo Noko University. Endo told Goldstein that he did not think that the dogs had tumors—the pathologists had misinterpreted what they saw. He believed that the dogs had large amounts of undigested drug in their intestines. Endo thought the abnormalities were just the side effect of having been given very large doses of the drug, which were about one hundred times greater than would be administered to humans. Goldstein had also seen odd structures inside cells treated with very high doses of compactin. Perhaps the concerns about the toxicity of compactin might have been overblown?

Goldstein and Brown were eager to find out whether reductase inhibitors might in fact work well in humans, particularly in those at greatest risk, such as FH patients. So they teamed with up two physicians, David Bilheimer and Scott Grundy, to test whether lovastatin could reduce LDL levels in a small trial of six FH patients with high levels of cholesterol and LDL. As they predicted, and hoped, treatment with lovastatin increased the number of LDL receptors and lowered LDL cholesterol levels by about 27 percent.

Excited by these results, Brown and Goldstein wrote to Vagelos at Merck. They explained how the drug treatment "reverses the genetic defect" of the deficiency of LDL receptors in FH patients and offered an entirely novel approach to treating genetic diseases. "Rather than replacing genes, we can exploit regulatory principles to make the good genes work harder," they said. It had been ten years since compactin had been discovered and three years since Merck and Sankyo had abandoned the development of reductase inhibitors. Goldstein and Brown urged Vagelos and Merck to resume working on them "as fast as possible."

Within a few months, Merck restarted larger trials of lovastatin but only in high-risk patients with extremely high cholesterol and cardiovascular disease. Executives remained concerned that lovastatin was carcinogenic or otherwise toxic. Merck's new head of Basic Research, Dr. Edward Skolnick, thought that lovastatin could be a very important drug if these concerns could be removed, so he assembled a team to undertake a comprehensive study to examine any potential toxicity. Skolnick's arrival on the project and championing of the drug was great news to Brown and Goldstein—the three men had known one another since their residencies at Massachusetts General Hospital. Goldstein and Skolnick later shared a lab at NIH and become good friends. Skolnick traveled to Texas to visit his former mates and to learn all he could about cholesterol regulation.

Goldstein and Brown suggested a clever experiment that could distinguish whether the lesions noted in animals were a direct effect of the drug, or might be explained by the large doses given previously and easily prevented. To Skolnick's delight, their suggestion worked like a charm, and no lesions appeared, nor did other tests indicate carcinogenic activity. Skolnick was confident that lovastatin was safe to use in humans.

After two years of testing, lovastatin was found to reduce plasma and LDL cholesterol levels by 20–40 percent. Merck applied for and received FDA approval to market the drug in August 1987.

But even with good clinical results and FDA approval, there was still substantial doubt among physicians about the general utility of the drug. After all, the goal was not the ability to reduce cholesterol levels but to reduce *deaths*. To examine the longer-term benefits of statins, Merck subsequently sponsored a five-year study of 4,444 patients using a next-generation statin (simvastatin, or Zocor). The results were better than anyone could have hoped: the study found a stunning 42 percent reduction in coronary deaths.

With such impressive benefits, the statin revolution swung into high gear. Thanks in considerable part to the use of these drugs, the death rate of Americans due to heart disease has declined by almost 60 percent since Ancel Keys raised the alarm about cholesterol.

It is a revolution that would not have happened without Brown's and Goldstein's discovery of the key rules of cholesterol regulation, Endo's original idea for searching for natural reductase inhibitors in fungi, and the perseverance of Merck's leadership and a small number of clinicians.

For their contributions, Brown and Goldstein shared the 1985 Nobel Prize in Physiology or Medicine. [Figure 4.2] Vagelos, who became CEO of Merck in 1985, led the company through a remarkable decade of innovation and commercial success.

And Endo? As it turned out, Endo never received a penny from his invention, and for a long time his contributions to the development of statins went unrecognized. This oversight was partly rectified when a symposium was held in his honor in Kyoto in 2003

FIGURE 4.2 Joseph Goldstein and Michael Brown. The photo was taken on the day of the announcement that they were to share the 1985 Nobel Prize in Physiology or Medicine.

Photo courtesy of Joseph Goldstein.

on the thirtieth anniversary of his discovery of compactin. In their tribute, Brown and Goldstein said, "Without Endo, the statins might never have been discovered. . . . The millions of people whose lives will be extended through statin therapy owe it all to Akira Endo and his search through fungal extracts."

CHAPTER 5

STUCK ACCELERATORS
AND BROKEN BRAKES

The motive that will conquer cancer will not be pity
or horror; it will be curiosity to know how and why.

—H. G. WELLS

Bicycles are ubiquitous around colleges, so students would not have noticed one particular red bike that rolled daily across the University of Chicago campus. But if they happened to look more closely, they would have been surprised by its elegant, gray-haired rider. Still, they never could have guessed that this happy commuter was eighty-eight years old, nor that this grandmother of five was one of the most decorated American scientists, having recently received the country's highest civilian honor, the Presidential Medal of Freedom.

But Janet Davison Rowley was just that. A pioneer in cancer research, Rowley was pivotal in establishing that cancer was a genetic disease. While the problem she was studying was much more complex than the regulation of sugar metabolism in bacteria or of cholesterol regulation in humans, her approach was the same as that taken by Monod and Jacob, and Brown and Goldstein, namely: find a situation where the rules of regulation have been broken and figure

FIGURE 5.1 Janet Rowley on her way to the laboratory.

Photograph by Dan Dry. Courtesy of the *University of Chicago Magazine*.

out what happened. Her breakthrough shaped a new view of cancer and ultimately led to a new kind of life-saving drug.

CHROMOSOMES AND PAPER DOLLS

Janet Davison grew up during the Depression in Chicago. Money was short, which required the family to move often from one neighborhood to another, and for young Janet to switch schools. The hard times precluded any luxuries or exotic hobbies. Her father introduced her to stamp collecting, and she became quite good at analyzing fine details. At an early age, she was able to pick out subtle differences among issues, such as whether there was a period or other distinguishing marks. She maintained the hobby well into adulthood, and the pattern recognition skills she acquired would turn out to serve her well.

After just two years of high school, Davison won a scholarship for a special program at the University of Chicago that enabled her to enroll at age fifteen and to complete both high school and a bachelor's degree. Davison thrived in the challenging but supportive atmosphere, and she developed a strong interest in biology and medicine. She then applied to and was accepted into the University's medical school. But in 1944, there was a quota for women, just three spots in a class of sixty-five students. The quota was filled, so Davison had to wait a year, which she did not see as a major setback, since she was still just twenty when she enrolled. She earned her medical degree in 1948, and the day after graduation married fellow medical student Donald Rowley.

While Davison (now Dr. Rowley) completed her education and training at an early age, she did not pursue a research career until much later. When she entered medical school, she looked forward to becoming a wife and mother, and she thought that being a physician would be an interesting, part-time profession. So, after completing their internships, she and Donald started their family. Rowley focused on raising four boys while working a few days a week at various clinics, first in Maryland and then back in Chicago.

One of those clinics served children with developmental disabilities. In 1959, only a couple of years after the correct total number of chromosomes in humans (forty-six) had been determined, it was discovered that children with Down syndrome had an extra copy of chromosome 21. Although she had never taken a genetics class, Rowley became fascinated with the inheritance of genetic diseases. After having worked in the clinic for several years, she wanted to do something more challenging.

The opportunity soon came when Donald arranged to spend his sabbatical year at Oxford University with Howard Florey, Charles Elton's former tentmate and co-recipient of the 1945 Nobel Prize for the development of penicillin. Rowley figured that she could use the year in England to learn how to analyze chromosomes and then bring that expertise back to the Chicago clinic. The technique at the time involved taking blood cells and growing them in the presence of a radioactive precursor of DNA, then imaging the radioactive chromosomes with sensitive film. The procedure allowed an accurate count of chromosomes and could detect gross abnormalities, but it was hard to discern details that distinguished different chromosomes.

Rowley learned the procedure well enough to be a co-author on a paper that analyzed how chromosomes were copied as cells multiplied. On returning to Chicago, she decided that she did not want to continue at the clinic but to pursue research instead. With just one paper to her credit, she approached Dr. Leon Jacobson, director of the Argonne Cancer Research Hospital and former chief physician to the Manhattan Project research team at the University of Chicago.

"I have a research project started in England that I'd like to continue with. Could I work here part time[?]" Rowley explained. "All I need is a microscope and a darkroom. And by the way, will you pay me? I must earn enough for a baby sitter."

Jacobson agreed, and Rowley began looking at the chromosomes with various blood disorders. In some patients' cells, she could tell that there were extra or missing chromosomes, but she could not determine the specific chromosomes involved. It was not until a new banding technique was invented using a fluorescent dye that those details became discernible. Rowley learned that technique on another sabbatical in England. When she returned, she started looking at more samples from leukemia patients.

In early 1972, she noticed something unusual in cells from two acute myeloid leukemia patients: pieces of two different chromosomes (numbers 8 and 21) appeared to have broken off and swapped places. Known as a *translocation*, the fact that the same swap had taken place in two different patients with the same type of cancer was striking.

At the time, it was well known that cancer cells often had abnormal numbers or kinds of chromosomes. But on close inspection, many cancers often appeared to be quite heterogeneous. Because there did not appear to be any consistent pattern, the chromosomal aberrations were generally considered a later consequence, not a cause, of the cancer. Indeed, the idea that cancers had specific genetic causes was not widely accepted. When Peyton Rous received the 1966 Nobel Prize for his discovery of a virus that caused cancer in chickens, he stated, "A favorite explanation has been that oncogens [cancer-causing agents] cause alterations in the genes of the cells of the body. . . . But numerous facts, when taken together, decisively exclude this supposition."

Rowley was excited. Perhaps the specific change she observed in the chromosomes of two leukemia patients could be the cause of their disease? She wrote a short report and submitted it to the *New England Journal of Medicine*, the leading medical journal. It was rejected. When she called to ask why, she was told that the finding was deemed "unimportant." She sent the paper instead to the obscure French journal *Annales de Génétique*, which published it.

Shortly thereafter, Rowley began to look at the cells of patients with a different cancer called chronic myelogenous leukemia (CML). Working at home on her "off" days from the lab, she studied the details of their chromosomes. She took pictures of the stained preparations, cut out the individual chromosomes, attached them to scrap paper, and spread them out on the family dining room table. The chromosomes occurred in pairs, many joined near their centers with two long "arms" pointing up and down. The boys teased their mom about her playing with paper dolls.

Many years earlier, two Philadelphia researchers had discovered that some CML patients' cancer cells had one unusually small chromosome 22, dubbed the Philadelphia chromosome. When Rowley looked carefully at CML cells using the newer staining techniques, she detected that chromosome 9 was also longer than usual in three

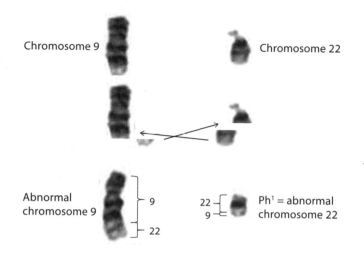

Chromosome 9

Chromosome 22

Abnormal
chromosome 9

9

22

22
9

Ph¹ = abnormal
chromosome 22

FIGURE 5.2 Chromosomal changes in a leukemia. Janet Davison Rowley noticed that not only was chromosome 22 abnormally small in chronic myelogenous leukemia cells (the Philadelphia chromosome), it had also exchanged ends with chromosome 9.

Reprinted by permission of Ruth MacKinnon.
http://scifundchallenge.org/firesidescience/2013/11/11/philadephia-the-birthplace-of-cancer-genetics/.

samples. In fact, the missing piece of chromosome 22 had been translocated to chromosome 9. This revealed that genetic information was not missing in CML cells, as previously thought: it had moved to a new location. [Figure 5.2]

With three independent cases of the chromosome 9;22 translocation in hand, she sent a report of her findings to *Nature*, the leading international science journal. The editors rejected the paper on the grounds that the translocation could just be a normal variation in the human population.

In the meantime, Rowley had examined the chromosomes of other cells in the patients' blood, which revealed forty-six normal chromosomes. The translocation was specific to the cancer cells. On top of that, she found several more independent cases of the same translocation among CML patients, nine in total. The anomaly could not be a coincidence. With this additional data, *Nature* was convinced and published the paper in the summer of 1973.

Rowley's discoveries of specific but different chromosomal changes in two different types of leukemia was strong evidence that at least some cancers were caused by specific, perhaps unique, genetic mutations. The translocations she had observed raised all sorts of questions. Were there others? How did they trigger cancer? The questions began to consume Rowley. Suddenly, her part-time job had become the most exciting part of her life. Contrary to her longstanding plan, and to her own surprise, her career and research moved from being merely a sideline to, at age forty-eight, being the center of her life. She started biking to the lab five days a week.

Soon thereafter, Rowley identified another translocation between chromosomes 15 and 17 in acute promyelocytic leukemia, and another group discovered a translocation in Burkitt's lymphoma.

A translocation meant that two segments of DNA that were not previously neighbors had become joined. Rowley suspected that the new juxtaposition of two genes must be the critical event in the onset of the cancers. But in the mid- to late 1970s, the human genome (the DNA of all twenty-three chromosomes that contains all our genes) was *terra incognita*. Identifying which genes were involved, let alone understanding how their translocation caused cancer, seemed far out of reach. Until, that is, some startling insights emerged from an altogether different line of investigation.

FINDING THE GENES FOR CANCER

If there was skepticism about the role of chromosomal changes in the origin of cancer, and there was plenty, the role of viruses had been doubted even more. In 1910, Rockefeller Institute scientist Peyton Rous had identified a virus capable of producing sarcomas in chickens. However, there was so much skepticism about his finding, and so little he could do beyond asserting the existence of the virus responsible, that he himself gave up the line of work. It was not until decades later—when the virus could be seen in powerful electron microscopes, and additional tumor-causing viruses were isolated from other animals—that the concept of cancer-causing viruses was firmly established. Hence, there was a fifty-six-year delay between Rous's discovery and his Nobel Prize. However, because no such viruses had been identified in humans, their role in human disease was at best unclear.

Still, the existence of tumor-causing viruses in animals offered potentially important clues to the rules for making cancer. One of the key advantages provided by studying a virus like the Rous sarcoma virus (RSV) was simplicity. RSV contained only a few genes, which raised the simple question: Which one(s) were involved in causing cancer?

The critical clue was discovered by Steve Martin, a graduate student at the University of California, Berkeley, who isolated a mutant RSV that was able to reproduce in cells but was unable to transform them into cancerous cells. The mutation responsible was in a single viral gene called *src* (pronounced "sark"), one of just four genes in the virus. Since an intact *src* gene was necessary for the virus to cause cancer, *src* was said to be a viral *oncogene* (a gene that induced cancer). But what could a chicken virus gene tell us about how cancer arose in general, particularly in humans?

The key insight from tumor viruses was obtained by another NIH-trained physician duo, Goldstein's and Brown's "classmate" Harold Varmus, and J. Michael Bishop. At NIH, Varmus studied the very enzyme regulation system pioneered by Monod and Jacob, while Bishop studied the polio virus. The two first teamed up when Varmus joined Bishop's laboratory at the University of California, San Francisco, in 1970 to study tumor viruses, RSV in particular; they later led a joint laboratory.

The discovery of the *src* gene provoked Varmus and Bishop to wonder about the origin of such a gene. The virus did not require it to infect or replicate in cells. If the virus did not need it to propagate itself, why did the gene exist, and where did it come from? Perhaps, they thought, it might be a cellular gene that the virus accidentally hijacked at some point in its history. If so, then one might find a *src* gene in normal chicken cells.

Looking for such a gene was much easier said than done in the era before the development of genetic engineering. It took nearly four years before the critical experiment could be performed in which a radioactively labeled DNA "probe" representing the viral src gene was used to search for a similar gene in the DNA of normal cells. The first revealing results were obtained by postdoctoral fellow Dominique Stehelin in October 1974: chicken cells indeed had a similar *src* gene. They called it c-*src* (for cellular *src*) to distinguish it from v-*src*, the viral oncogene. C-*src* was soon found in other birds, including ducks and turkeys, even an emu.

But c-*src* was not just a bird gene. Varmus, Bishop, and their colleague Deborah Spector found the gene in mammals as well, including humans. Its presence in different animals revealed that c-*src* was an old gene that had existed for several hundred million years.

What could this mean? Varmus and Bishop pondered the exciting possibilities. First, the long evolutionary pedigree of c-*src* indicated that it had some job in normal cells. And, second, the close similarity between v-*src* and c-*src* suggested that the tumor-causing RSV could have acquired its v-*src* gene by hijacking a copy of the c-*src* gene, and in doing so, altered the gene so that it now caused cancerous growth.

Still, *src* was just one case. Were there other tumor virus oncogenes that had cellular counterparts? The race was on to find them. Very quickly, several viral oncogenes were discovered in different tumor viruses that infected chickens, mice, or rats, and their counterparts were found to exist not just in their host species, but in other animals, including humans. Genes named *myc*, *abl*, and *ras* extended the paradigm that viral oncogenes came from normal cellular genes, called proto-oncogenes.

By the late 1970s, there appeared to be two strong but unconnected bodies of evidence emerging as to the origin of cancer. The existence of viral oncogenes and cellular proto-oncogenes offered a powerful explanation of how viruses could cause cancers, but only in animals. The demonstration of consistent chromosomal changes in some human cancers was compelling but was limited to certain tumor types, and the genes involved were unknown. Was there any way to connect the two?

There was, and that connection would reveal how cancer was a disease of regulation.

BREAKING THE RULES OF REGULATION

Among the first handful of viral oncogenes and cellular proto-oncogenes discovered after *src* was the v-*abl* gene of the mouse Abelson leukemia virus and its cellular counterpart, the c-*abl* gene. Like c-*src* and the rest, c-*abl* also exists in the human genome. When researchers found that the c-*abl* gene mapped to human chromosome 9, the same chromosome implicated in the translocation Rowley described in CML, they wondered: Could there be some connection? Was the chromosome broken near the c-*abl* gene in CML patients?

It was a long shot. Chromsomes are very large, each containing an average of about 1,000 genes. C-*abl* could have been anywhere on the chromosome. A team of Dutch and British researchers probed cells containing the Philadelphia chromosome (number 22), and they were astonished to find that the c-*abl* gene from chromosome 9 had in fact moved to chromosome 22 (see Figure 5.3, left).

This discovery raised the exciting possibility that c-*abl* could be directly involved in the human cancer. To find out more about what happened to c-*abl* in CML cells, the researchers isolated the part of chromosome 22 that had been juxtaposed to the c-*abl* gene. Remarkably, it turned out that the c-*abl* gene had moved to almost the same place on chromosome 22 in seventeen different patients. So not only was the translocation from chromosome 9 to 22 a hallmark of the cancer, the chromosomes were joined at essentially the same location. This suggested that there was something very important about where c-*abl* was joined to chromosome 22. Further examination revealed that the c-*abl* gene was fused to another gene called *bcr* (for breakpoint cluster region). As a result, the fused gene produced an abnormal protein in which the "head" of the c-abl protein was fused to the "tail" of the bcr protein (see Figure 5.3, right).

Somehow, this fusion turned a normal proto-oncogene into a deadly oncogene. How it did so became clear when researchers compared the activities of the normal c-abl protein and the bcr/abl fusion protein. The c-abl protein is a member of class of enzymes called tyrosine kinases that work by adding phosphate groups to proteins. The addition and removal of phosphates is another common way

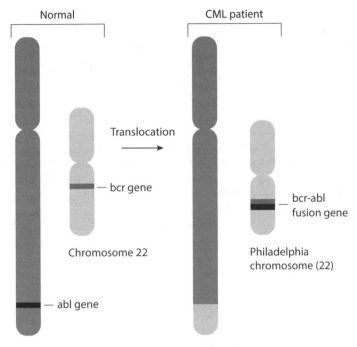

FIGURE 5.3 The fusion of two genes creates a cancer-causing oncogene. In chronic myelogenous leukemia cells the *abl* gene from chromosome 9 is fused to the *bcr* gene on chromosome 22; the hybrid gene produces a protein with abnormally elevated activity.

Illustration by Leanne Olds.

that protein activities are regulated, by flipping them between active and inactive states. Many kinases are parts of chemical relay systems that transfer information from outside cells to the internal machinery that determines whether the cell multiplies, differentiates, or dies. The activity of the c-abl tyrosine kinase activity is usually low in normal cells. But the activity of the bcr/abl fusion protein is much higher. The fusion created a mutant enzyme that, like the constitutive mutants studied by Monod and Jacob was always "on."

Leukemia, then, is a disease of regulation. In CML, the normal control of the multiplication of white cells is broken by the mutant bcr/abl protein. The hyperactive protein interferes with multiple relay systems in cells such that the signals to multiply are stuck in the "on" state, like a stuck accelerator in a car. As it turns out, mutations in several dozen other oncogenes implicated in myriad other cancers appear to have the same general effect, revealing that cancer in general is a disease of regulation.

While the discovery of oncogenes and their modes of action was a leap forward in understanding cancer, it turns out that oncogenes are just half of the genetic story of cancer. And perhaps by this point in this book, with what we have seen about regulatory logic, you might be able to guess the nature of the other half. A stuck accelerator is certainly not the only way for a car to careen out of control. What might be another mechanism? (Hint: think about negative regulation and double-negative regulatory logic.)

The same effect can result from a foot slipping off the brake, or cutting the brake lines. Researchers have found that indeed the loss of genetic "brakes" is a very common event in the genesis of cancer.

TUMOR SUPPRESSORS

The first genetic "brake" discovered involved a rare cancer of the eye called retinoblastoma. The disease usually develops in children at an early age and sometimes runs in families. The crucial clue to the genetic mystery of retinoblastoma was that in some cases, a portion of both copies of chromosome 13 was deleted, suggesting that the loss of both copies of some gene was the key event in the formation of retinoblastoma. This situation is in contrast to oncogenes, where the alteration of one copy of the gene (e.g., *bcr/abl*) was the key event in cancer formation.

In genetic parlance, we say that oncogene mutations are dominant, because they exert their effects even if the normal proto-oncogene is intact. In contrast, the retinoblastoma mutation is recessive, because both copies have to be altered for disease to form. It appears then that the normal function of the missing gene is necessary to prevent or suppress tumor formation, so this type of gene was dubbed a "tumor suppressor."

By zeroing in on the DNA that was missing in retinoblastoma patients, the retinoblastoma gene (dubbed *Rb*) was identified. *Rb*'s function, of course, is not to produce cancer, that is a consequence of its loss or alteration. Extensive studies of the Rb protein have revealed that its normal job is to control a critical decision in the life cycle of cells. For cells to multiply, they must copy their DNA and then divide into two. This process is highly regulated and progresses in phases. Rb acts at a major early checkpoint in this cycle by blocking the decision to replicate DNA. Loss of both copies of the *Rb* gene, then, allows cells to continue replicating in an uncontrolled way.

Rb is not the only tumor suppressor gene; about seventy such genes have now been identified. Nor are *Rb* mutations associated only with retinoblastoma; *Rb* mutations are found in other cancers, such as osteosarcoma and lung cancers.

Moreover, mutations are not the only way to inactivate Rb. The activity of the Rb protein is regulated by the addition of phosphate groups by protein kinases: Rb is most active when it is less phosphorylated, and inactivated by heavy phosphorylation. The direct or indirect effect of many oncogene proteins, including bcr/abl, is to increase the phosphorylation of Rb, which inhibits Rb activity and allows for continuous cell multiplication. Indeed, Rb is inactivated in some way in most, perhaps all, human cancers.

Here again is the sort of negative regulation and double-negative regulatory logic we have encountered before. Generally speaking, Rb is a repressor of cell multiplication. Cell growth, then, normally requires inhibition of this repressor to proceed. But inactivation (left) or loss (right) of Rb allows for constitutive growth:

Cancer via oncogene mutation	*Cancer via tumor suppressor mutation*
Oncoprotein (bcr-abl)	Oncoproteins
⊥	⊥
Rb protein inactive	Rb gene lost
⊥	⊥
DNA replication,	**DNA replication,**
Cell multiplication	**Cell multiplication**

The role of Rb fits remarkably well with Monod's and Jacob's speculation many decades ago about cancer resulting from the inactivation of a repressor of cell multiplication (see Chapter 3).

With this knowledge of how mutations in certain genes break the rules of the regulation of cell growth, the great challenge is to figure out how, if possible, to put the brakes back on in cancer cells.

LOGICAL THERAPY AND RATIONAL DRUGS

For many decades, cancer has been treated with surgery, irradiation, and cocktails of drugs that kill dividing cells. The latter are blunt instruments that do not target cancer cells specifically, so they are hampered by varying effectiveness, and cause many debilitating and dangerous side effects. The long-held hope of cancer research has been to design more effective, safer therapies that are tailored to a patient's specific cancer. That hope has become a reality. The pioneer in this new class of drugs, called Gleevec, targets the very mutation that Janet Rowley spotted on her dining room table.

But like many pioneers, Gleevec almost died several times before reaching its destination. Indeed, the parallels between the story of Gleevec and the development of the first statin are uncanny. But thanks again to a physician who saw the need and was tireless in encouraging drug development, there was spectacular clinical success that changed medical history.

The *bcr/abl* translocation creates a hyperactive protein kinase that causes the inactivation of the *Rb* repressor, allowing uncontrolled cell multiplication. What was needed was something that addressed the double-negative logic of CML—something that would inhibit *bcr/abl* and block the renegade enzyme from doing its harm.

Nick Lydon and Alex Matter, two scientists at the Ciba-Geigy pharmaceutical company in Basel, Switzerland, figured that since many oncogenes produced altered kinases, inhibitors of the enzymes might block cancer cell growth. Rather than searching through nature, as Endo did, or using the traditional trial and error methods of the pharmaceutical industry, they used a method called "rational design" to devise molecules that were to fit into and block the active site of kinases, so that the normal "key" can't fit into the "lock." After years of chemistry and testing, they had several candidate compounds, including a molecule that inhibited the normal c-abl kinase.

To find out whether any of their compounds could work on CML cells, Lydon offered them to a physician he knew. Brian Druker of the Oregon Health Sciences University in Portland was not only very interested in potential inhibitors of the bcr/abl kinase but, very importantly, he had access to CML patients' cells. Druker found that one particular compound Lydon gave him killed CML cells, but not normal cells, at very low concentrations.

While Druker, Lydon, and Matter were all excited by the results, the company did not think there was much of a market for a CML-specific drug. It took more than a year to convince them to advance to further testing in animals. Then, the first toxicology tests in dogs raised concerns that the drug was unsafe for intravenous use in humans. Soon thereafter, Ciba-Geigy merged with the Sandoz pharmaceutical company to create the new company Novartis. After the merger, the drug languished, and Lydon resigned.

Novartis scientists eventually put an oral formulation of the drug through animal testing, but the results again raised concerns. One toxicologist told Matter, "Not over my dead body will this compound go into man."

Druker, however, was not deterred. His patients' prognoses were grim—25 to 50 percent died in the first year after diagnosis—and the most he could do with available therapies was to buy time. Druker thought that any drug toxicity could be managed by monitoring patients and modifying the dosage. He urged Matter to "give the drug a chance." Matter kept pressing company management on the need for the drug. Finally, the new CEO of Novartis, Daniel Vasella, backed a clinical trial in humans. The study began in June 1998, almost five years after Druker had first tested the drug on CML cells in the laboratory.

Druker and two other physicians started administering the drug to a small number of CML patients, gradually increasing the dosage while monitoring both their disease and potential side effects. The main indicator of any benefit would be a drop in white cell count: normally there are about 4,000–10,000 cells per microliter of blood; in CML patients this can rocket to 100,000–500,000 cells per microliter. At low doses, they saw no effect. Then, as they administered larger doses, some patients' white cell counts dropped back toward normal. Inspection of their blood under the microscope revealed that the fraction of cells bearing the Philadelphia chromosome was also dropping. The drug was killing its target.

Novartis put its full resources behind the drug. The trial was expanded, the dosages were increased, and the patients followed for several months. Ninety-seven percent of patients who received the highest dose had their white blood counts return to normal, usually in just four to six weeks. In three-quarters of patients, the Philadelphia chromosome–containing cancer cells vanished. The results weren't good—they were great, unprecedented in the history of chemotherapy. The FDA gave the drug priority review and approved it in less than three months in May 2001.

Thanks to Gleevec, the prognosis for CML patients shifted dramatically. Long-term survival rates (longer than eight years) jumped to nearly 90 percent compared to about 45 percent before the introduction of the drug. Contrary to company forecasts, the drug turned out to be a blockbuster for Novartis, generating almost $28 billion in sales over ten years. In 2012, Lydon, Druker, and Rowley shared the prestigious Japan Prize for their respective contributions to the understanding and treatment of CML.

The rational approach to CML therapy was a spectacular success. But Gleevec is just one drug, *bcr/abl* is just one oncogene, and CML is just one type of cancer. What can we say or do about other cancers?

KNOW THY ENEMY—AND KILL IT

There are more than two hundred different cell types among the 37 trillion cells that make up the adult human body. Building and maintaining the right number of so many different kinds of cells requires a lot of regulation. It also requires the copying of many trillions of long molecules of DNA. Mistakes get made in the copying of DNA. Most of these mutations are harmless, but some create the potential for disaster. Understanding which specific mutations have occurred in an individual cancer is key to a more precise diagnosis and to targeted therapy.

The technology for analyzing cancers has come a long way since Janet Rowley cut out pictures of chromosomes and arranged them on her dining room table. Because of advances that have increased the speed and reduced the cost of DNA sequencing, it has become possible to peer into any tumor and assess the integrity of every gene.

By studying thousands of tumors of all sorts of tissues, researchers have built a catalog of gene mutations. By looking for genes that are frequently mutated (or lost) in each cancer, most of the genes that commonly contribute to cancer have been identified.

One major revelation from these surveys is that only a small fraction of all human genes are implicated in cancer. Of the roughly 20,000 genes in the human genome, about 140 have been found that are frequently mutated, with an almost even split between oncogenes and tumor suppressors. That figure can be seen as generally good news for researchers, doctors, and patients, as it narrows the number of players that we need to know about. And better still, we know a lot about the normal roles of these genes: virtually all of them are components of just a dozen well-studied relay systems or pathways that regulate cell differentiation or survival.

The surveys also reveal that most cancers contain mutations in two to eight of the 140 genes. Knowing which genes are altered in individual tumors provides a new way to classify them according to their genetic makeup, to relate the mutations to cancer behavior, and to target therapy against those mutations. In 1997, there were no approved drugs that specifically targeted mutations in any cancer; in 2015, there are more than three dozen drugs, with many more in the research pipeline. We are gaining ground, but it is far too early to declare victory.

In 2010, Janet Rowley was diagnosed with ovarian cancer. Throughout her treatment, she had biopsies and other samples of the tumor sent to her colleagues for study. But on December 17, 2013, she passed away from complications. She had pre-arranged her own autopsy, so that researchers could learn how her disease progressed.

I opened this chapter with a quote from H. G. Wells about the motives that would lead to the conquering of cancer. I omitted the preceding sentence of dialogue from the character in his novel *Meanwhile*: "The disease of cancer will be banished from life by calm, unhurrying, persistent men and women, working, with every shiver of feeling controlled and suppressed, in hospitals and laboratories."

The scientists who have made the greatest strides against cancer have neither been unhurrying, nor have they controlled or suppressed their feelings of urgency and compassion.

PART III
THE SERENGETI RULES

The regulation of populations must be known before
we can understand nature and predict its behavior.

—NELSON HAIRSTON, FREDERICK SMITH, AND LAWRENCE SLOBODKIN (1960)

We have explored the rules and logic that regulate the numbers of certain molecules and cells in the body, and we have seen the catastrophic consequences when they are broken. We have also seen how intimate knowledge of specific rules can be used to cure the sick. I now turn to the rules of regulation on the much larger scale of animal and plant populations, and how knowledge of those rules can be used to heal ailing species and habitats.

The central question was posed by Charles Elton (Chapter 2): How are the numbers and kinds of animals and plants regulated?

Take, for example, the great Serengeti. It is home to an astounding number of different kinds of animals—over seventy species of mammals, more than 500 species of birds, and even one hundred different species of dung beetles. Among those mammals are some of the rarest (wild dogs), fastest (cheetah), largest (the African elephant) and most numerous (wildebeest) animals.

What rules regulate the numbers of such different creatures?

While there is no better place than the Serengeti to appreciate the magnitude of the question, there are easier places to figure out some of the basic rules. The art of biology is to find the simplest model of the phenomena one wants to understand—as the pioneers in the previous stories did with an enzyme or a tumor virus—and to conduct carefully controlled experiments that manipulate one variable at time. The richness of the vast, intact Serengeti is also a scientific handicap in that it is pretty difficult (but not impossible) to do controlled experiments on migrating wildebeest, lion prides, and elephant herds!

Just as for the specific rules governing cholesterol and cell growth, there are two good ways to discover the rules that regulate animal populations: find systems where the rules can be broken experimentally and see what happens, or find situations where the system (in this case, an ecosystem) is broken and decipher why.

To identify some of the rules that regulate populations, I will first look at some pioneering discoveries in various parts of the world (Chapter 6), then explore how they and a few additional rules operate in the great Serengeti (Chapter 7). I will then turn to some places where these rules have been broken (Chapter 8), and to some extraordinary efforts to restore entire ecosystems (Chapters 9 and 10).

What these pioneers discovered was a set of ecological rules that are surprisingly analogous to the physiological rules I have described. Indeed, I deliberately introduced you first to negative and positive regulation, double-negative logic, and feedback regulation at the molecular scale. You are about to see their power on the grandest stage.

CHAPTER 6

SOME ANIMALS ARE MORE EQUAL THAN OTHERS

You push an ecological system too far and
suddenly all the rules change.

—ROBERT PAINE

Even in 1963, one had to go pretty far to find places in the United States that were not disturbed by people. After a good deal of searching, Robert Paine, a newly appointed assistant professor of zoology at the University of Washington in Seattle, found a great prospect at the far northwestern corner of the lower forty-eight states. On a field trip with students to the Pacific coast, Paine wound up at Mukkaw Bay, at the tip of the Olympic Peninsula. The curved bay's sand and gravel beach faced west onto the open ocean and was dotted with large outcrops. Among the rocks, Paine discovered a thriving community. The tide pools were full of colorful creatures—green anemones, purple sea urchins, pink seaweed, bright red Pacific blood starfish, as well as sponges, limpets, and chitons. Along the rock faces, the low tide exposed bands of small acorn barnacles, and large, stalked goose barnacles; beds of black California mussels; and some very large, purple and orange starfish, called *Pisaster ochraceus*.

"Wow, this is what I have been looking for," he thought.

FIGURE 6.1 The ochre starfish Pisaster ochraceus in the rocky intertidal zone on the Pacific coast. The starfish prey on the mussels, which enables other species such as kelp and small animals to co-occupy the community.

Photo courtesy of David Cowles, rosario.wallawalla.edu/inverts.

The next month, June 1963, he made the four-hour journey back to Mukkaw from Seattle, first crossing Puget Sound by ferry, then driving along the coastline of the Strait of Juan de Fuca, onto the lands of the Makah Nation, and out to the cove of Mukkaw Bay. At low tide, he scampered on to a rocky outcrop. Then, with a crowbar in hand and mustering all of the leverage he could with his six feet six inch frame, he pried loose every purple or orange starfish on the slab, grabbed them, and hurled them as far as he could out into the Bay.

So began one of the most important experiments in the history of ecology.

WHY IS THE WORLD GREEN?

Paine's journey to Mukkaw Bay and its starfish was a circuitous one. Born and raised in Cambridge, Massachusetts, Paine was named for his ancestor Robert Treat Paine, who was a signer of the Declaration of Independence. His interests in nature were fueled by exploring the New England woods. His first love was bird-watching, with butterflies and salamanders close seconds.

Paine went on frequent bird-watching walks with a neighbor, who insisted that he make a record of all his sightings. It was good discipline, for Paine became such a sharp bird-watcher that he became the youngest member of the Nuttall Ornithological Club, an exclusive tribe of top birders.

He was also inspired by the writings of prominent naturalists, who opened his eyes to the drama of wildlife. Passages such as this from Edward Forbush's *Birds of Massachusetts* stoked his youthful imagination:

A gang of men were startled one winter's day ago in Medfield woods by the sight of what seemed like a huge bird "with four wings" whirring past. The "bird" dropped into the snow not far away. It turned out to be a Goshawk and a Barred Owl locked in deadly clasp. Both were dead when picked up.

He was just as enthralled by intimate accounts of spider behavior as by Jim Corbett's hair-raising tales of tracking down tigers and leopards in rural India in *Man-eaters of Kumoan*. Spiders, in fact, were considered "sacred objects" in the Paine household. Young Paine spent many hours admiring the orb weavers preying on houseflies he provided.

After enrolling at Harvard, and inspired by several famous paleontologists on the faculty, Paine developed an intense new interest in animal fossils. He was so fascinated by the marine animals that lived in the seas more than 400 million years ago that he decided to study geology and paleontology in graduate school at the University of Michigan.

The course requirements entailed rather dry surveys of various animal "ologies"—ichthyology (fishes), herpetology (reptiles and amphibians), and so forth that Paine found boring. One exception was a course on the natural history of freshwater invertebrates taught by ecologist Fred Smith. Paine appreciated how the professor provoked his students to think.

One memorable spring day, the sort of day when professors don't feel like teaching and students don't want to be inside, Smith told the class, "We are going to stay in this room." He looked outside at a tree that was just getting its leaves.

"Why is that tree green?" Smith asked, looking out the window.

"Chlorophyll," a student replied, correctly naming the leaf pigment, but Smith was heading down a different path.

"Why isn't all of its greenery eaten?" Smith continued. It was such a simple question, but Smith showed how even such basic things were not known. "There is a host of insects out there. Maybe something is controlling them?" he mused.

At the end of his first year, Smith sensed Paine's unhappiness with geology, and suggested that he consider ecology instead. "Why don't you be my student?" he asked. It was a major change in direction, and there was a catch. Paine proposed to study some fossil animals from the Devonian period in nearby rocks. Smith said, "No way." Paine had to study living, not extinct creatures. Paine agreed, and Smith became his adviser.

Smith had long been interested in brachiopods or "lamp shells," marine animals with an upper and lower shell, joined at a hinge. Paine knew about the animals because they were abundant in the fossil record, but their present-day ecology was not well known. Paine's first task was to find living forms. Lacking a nearby ocean, Paine made scouting trips to Florida in 1957 and 1958, and found some promising locations. With Smith's approval, he began what he called his "graduate student sabbatical." In June 1959, he drove back to Florida and began living out of his Volkswagen van. For eleven months he studied the range, habitat, and behavior of one species.

It was the sort of work that provided a solid foundation for a naturalist-in-training, and it would earn Paine his PhD. But the filter-feeding brachiopods were not the most dynamic animals. And sifting large amounts of sand for the less than quarter-inch-long creatures was, well, just not very exciting.

As Paine shoveled his way along the Gulf Coast, it was not Florida's brachiopods that captured his imagination. On the Florida panhandle, Paine discovered the Alligator Harbor Marine Laboratory and was given permission to stay there. At the tip of nearby Alligator Point, he noticed that for a few days each month, the low tide exposed an enormous gathering of large predatory snails—some, such as the horse conch, more than a foot long. The mud and saw-grass of Alligator Point was not at all boring, quite the contrary—it was a battlefield.

On top of his thesis work on brachiopods, Paine made a careful, Eltonian study of the snails. He counted eight abundant snail species, and took detailed notes on who ate whom. In this "gastropod eats gastropod" arena, Paine saw that without exception, it was always a larger snail devouring a smaller one, but not everything that was smaller. The eleven-pound horse conch, for example, dined almost exclusively on other snails, and paid little attention to smaller prey, such as the clams that were the main fare for the smaller snails. The junior scientist interpreted his data in Eltonian terms:

> Elton (1927) suggested that size relationships were the chief reason for the existence of food chains since what is too large, or too small, to be consumed by one organism can be eaten by another. In this way, small items after passing through one or more intermediate steps become indirectly available to larger predators.

While Paine was in Florida watching predators up close, his advisor Smith had kept thinking about those green trees and the roles of predators in nature. Smith was keenly interested not just in the structure of communities, but also in the processes that shaped them. He often had bag lunches with two colleagues, Nelson Hairston Sr. and Lawrence Slobodkin, during which they had friendly arguments about major ideas in ecology. All three scientists were interested in the processes that control animal populations, and they debated explanations circulating at the time. One major school of thought was that population size was controlled by physical conditions, such as the weather. Smith, Hairston, and Slobodkin (hereafter dubbed "HSS") all doubted this idea, because, if true, it meant that population sizes fluctuated randomly with the weather. Instead, the trio was convinced that biological processes must control the abundance of species in nature, at least to some degree.

Like Elton's pyramid, HSS pictured the food chain as subdivided into different levels according to the food each consumed (known as trophic levels). At the bottom were the decomposers that degrade organic debris; above them were the producers, the plants that relied on sunlight, rain, and soil nutrients; at the next level were the consumers, the herbivores that ate plants; and above them the predators that ate the herbivores. [Figure 6.2]

The ecological community generally accepted that each level limited the next higher level; that is, populations were positively regulated from the bottom up. But Smith and his lunch buddies pondered the observation that seemed at odds with this view: the terrestrial world is green. They knew that herbivores generally do not completely consume all vegetation available. Indeed, most plant leaves only show signs of being partially eaten. To HSS, that meant that herbivores were not food-limited, and that something else was limiting herbivore populations. That something, they believed, were predators, negatively regulating herbivore populations from the top down in the food chain. While predator-prey relationships had long been studied by ecologists, it was generally thought that the availability of prey regulated predator numbers and not vice versa. The proposal that predators as a whole acted to regulate prey populations was a radical twist.

To bolster their case, HSS noted instances where herbivore populations had exploded after the removal of predators, such as the Kaibab deer population in northern Arizona that increased after decimation of local wolf and coyote populations. They assembled their observations and arguments in a paper titled "Community Structure, Population Control, and Competition" and submitted it to the journal *Ecology* in May 1959. It was rejected. The article did not see the light of day until the year-end issue of *American Naturalist* in 1960.

The proposal that predators regulate herbivore populations is now widely known as the "HSS hypothesis" or "Green World Hypothesis." While HSS declared, "The logic used is not easily refuted," their ideas, like most that challenge the status quo, drew a lot of criticism. I need not recount all of it here. One legitimate critique was that their claims needed testing and more evidence. And that was just what Smith's former student set out to do on Mukkaw Bay in 1963.

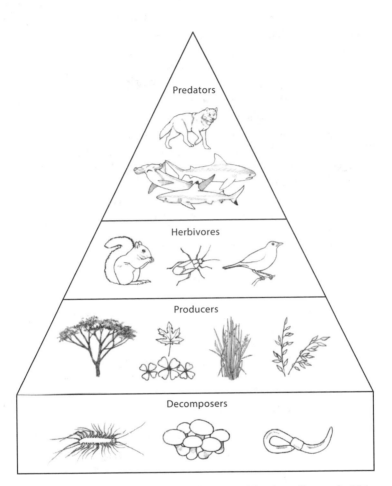

FIGURE 6.2 The trophic levels of biological communities. According to the Hairston, Smith and Slobodkin (HSS) green world hypothesis, organisms belong to one of four trophic levels: decomposers (fungi and worms), producers (plants and algae), herbivores, and carnivores.

Illustration by Leanne Olds.

KICK IT AND SEE

The HSS hypothesis was essentially a description of the natural world based on observation. So, too, were Elton's work and ideas, and Paine's own studies to date on brachiopods and predatory snails (as were Darwin's, for that matter). Indeed, virtually all of ecology up to the 1960s had been based on observation. The limitation of such observational biology was that it left itself open to alternative explanations and hypotheses. Paine, like the pantheon of molecular biologists I have described, realized that if he wanted to understand how nature worked—the rules that regulated animal populations—he would have to find situations where he could intervene and break them. In the specific case of the roles of predators, he needed a setting where he could remove predators and see what happened—what would later be described as "kick it and see" ecology. Hence, the starfish hurling.

Twice a month every spring and summer, and once a month in the winter, Paine kept returning to Mukkaw to repeat his starfish-throwing ritual. On one twenty-five-foot-long by six-foot-tall stretch of rock, he removed all the starfish. On an adjacent stretch, he let nature take its course. On each plot, he counted the number and calculated the density of the inhabitants, tracking fifteen species in all.

To understand the structure of the Mukkaw food web, Paine paid close attention to what the predators were eating. The starfish has the neat trick of everting its stomach to consume prey, so to see what it was feasting on, Paine turned more than 1,000 starfish over and examined the animals held against their stomachs. He discovered that the starfish was an opportunistic gourmand that ate barnacles, chitons, limpets, snails, and mussels. While the small barnacles were the most numerous prey—the starfish was able to scarf up dozens of the little crustaceans at a time—they were not its primary source of calories. Mussels and chitons were the most important contributors to the starfish diet.

By September, just three months after he began removing the starfish, Paine could already see that the community was changing. The acorn barnacles had spread out to occupy 60–80 percent of the available space. But by June of 1964, a year into the experiment, the acorn barnacles were in turn being crowded out by small but rapidly growing goose barnacles and mussels. Moreover, four species of algae had largely disappeared, and the two limpet and two chiton species had abandoned the plot. While not preyed on by the starfish, the anemone and sponges populations had also decreased. However, the population of one small predatory snail, *Thais emarginata*, increased ten- to twentyfold. Altogether, the removal of the predatory starfish had quickly reduced the diversity of the intertidal community from the original fifteen species to eight.

The results of this simple experiment were astonishing. They showed that one predator could control the composition of species in a community through its prey—affecting both animals it ate as well as animals and plants that it did not eat.

As Paine continued the experiment over the next five years, the line of mussels advanced down the rock face by an average of almost three feet toward the low tide mark, monopolizing most of the available space and pushing all other species out completely. Paine realized that the starfish exerted their strong effects primarily by keeping the mussels in check. For the animals and algae of the intertidal zone, the important resource was real estate—space on the rocks. The mussels were strong competitors for that space, and without the starfish, they took over and forced other species out. The predator stabilized the community by *negatively regulating* the population of the competitively dominant species. Schematically, the plots with and without starfish looked like:

Starfish			Starfish		
	⊥			⊥	
Barnacles ⇔	Mussels ⇔	Chiton	Barnacles ⇔	**Mussels** ⇔	Chiton
Limpets		Algae	Limpets		Algae

Paine's starfish-tossing was strong confirmation of the HSS hypothesis that predators exerted control from the top down. But this was just one experiment with one predator in one spot on the Pacific coast. If Paine was going to draw any generalities, it was important to test other sites and other predators. The dramatic results of the Mukkaw Bay experiments inspired a flurry of kick-it-and-see experiments.

Paine discovered uninhabited Tatoosh Island when he was out on a salmon fishing trip. On this small, storm-battered island, several miles up the coast from Mukkaw Bay and about half a mile offshore, Paine found many of the same species clinging to the rocks, including large *Pisaster* starfish. With the permission of the Makah tribe, Paine started tossing them back in the water. Within a few months, the mussels started spreading across the predator-free rocks.

While on sabbatical in New Zealand, Paine investigated another intertidal community at the north end of a beach near Auckland. There, he found a different starfish species called *Stichaster australis* that preyed on the New Zealand green-lipped mussel, the same species exported to restaurants around the world. Over a period of nine months Paine removed all starfish from one 400-square-foot area and left an adjacent, similar plot alone. He saw immediate and striking effects. The treated area quickly began to be dominated by mussels, which extended their range by about 40 percent down toward the low tide mark. Six of twenty other species initially present vanished in just eight months; within fifteen months the majority of space was occupied solely by the mussels. Interestingly, this expansion occurred despite the abundance of another large mussel predator (a sea snail).

To Paine, the predatory starfish of Washington and New Zealand were "keystones" in the structure of intertidal communities. Just as the stone at the apex of an arch is necessary for the stability of the structure, these apex predators at the top of the food web are critical to the diversity of an ecosystem. Dislodge them, and as Paine showed, the community falls apart. Paine's pioneering experiments, and his coining of the term "keystone species" prompted the search for keystones in other communities and would lead him to another seminal idea.

CASCADING EFFECTS AND DOUBLE-NEGATIVE LOGIC IN FOOD CHAINS

Paine's kick-it-and-see experiments were not limited to manipulating predators. He was interested in understanding the rules that determined the overall makeup of coastal communities. Other prominent inhabitants of the tide pools and shallow waters included a great variety of algae, such as the large brown seaweed known as kelp. But their distribution was patchy—abundant and diverse in some places, nearly absent from others. One of the most prevalent grazers on the algae were sea urchins. Paine and Robert Vadas set out to find out what effect the urchins had on algal diversity.

To do so, they removed all the urchins by hand from some pools around Mukkaw Bay, or barred them from areas within Friday Harbor (near Bellingham) with wire cages. They left nearby pools and areas untouched as controls for their experiment. They observed dramatic effects of removing the sea urchins—several species of algae burst forth in the urchin-free zones. The control areas with large urchin populations contained very few algae.

Paine also noticed that such urchin-dominated "barrens" were common in pools around Tatoosh Island. At face value, the urchin barrens seemed to violate a key assertion of the HSS hypothesis that herbivores tended not to consume all vegetation available. But the explanation for why there were such barrens in Pacific waters would soon become clear—in the surprising discovery of another keystone species, an animal that had been removed from Washington's coast long before Paine started tinkering with nature.

Sea otters once ranged from northern Japan to the Aleutian Islands and along the North American Pacific coast as far south as Baja California. Coveted for their luxurious fur, the densest of all marine mammals, the animals were hunted so intensively in the eighteenth and nineteenth centuries that by the early 1900s, only 2,000 or so animals remained of an original population of 150,000–300,000, and the species had disappeared from most of its range, including Washington state. The species gained protected status in 1911 under the terms of an international treaty. After their near-extermination from the Aleutian Islands, the animals rebounded to high densities in some locations.

In 1971, Paine was offered a trip to one of those places—Amchitka Island, a treeless island in the western part of the Aleutians. Some students were working on the kelp communities there, and Paine flew out to offer his advice. Jim Estes, a student from the University of Arizona, met with Paine and described his research plans. Estes was interested in sea otters, but he was not an ecologist. He explained to Paine that he was thinking about studying how the kelp forests supported the thriving sea otter populations.

"Jim, you are asking the wrong questions," Paine told him. "You want to look at the three trophic levels: sea otters eat urchins, sea urchins eat kelp."

Estes had only seen Amchitka with its abundant otters and kelp forests. He quickly realized the opportunity to compare islands with and without otters. With fellow student John Palmisano, Estes traveled to Shemya Island, a six-square-mile chunk of rock 200 miles to the west without otters. Their first hint that something was very different was when they walked down to the beach and saw huge sea urchin carcasses. But the real shock came when Estes dove under the water for the first time—and saw that the bottom was carpeted with large sea urchins and had no kelp. He saw other striking differences between the two communities around each island: colorful rock greenling, harbor seals, and bald eagles were abundant around Amchitka but not around otter-less Shemya.

Estes and Palmisano proposed that the vast differences between the two communities were driven by sea otters, which were voracious predators of sea urchins. They suggested that sea otters were keystone species whose negative regulation of sea urchin populations was key to the structure and diversity of the coastal marine community.

Estes's and Palmisano's observations suggested that the reintroduction of sea otters would lead to a dramatic restructuring of coastal ecosystems. Shortly after their pioneering study, the opportunity arose to test the impact of sea otters as they spread along the Alaskan coast and recolonized various communities. In 1975, sea otters were absent from Deer Harbor in southeast Alaska. But by 1978, the animals had established themselves there; sea urchins were small and scarce; the sea bottom was littered with their remains; and tall, dense stands of kelp had sprung up.

The presence of the otters had suppressed the urchins, which had otherwise suppressed the growth of kelp. Schematically, the rules of sea urchin and kelp regulation looked like:

Otters

$\perp$

Sea urchins

$\perp$

Kelp

Double-negative logic once again. In this instance, otters "induce" the growth of kelp by repressing the population of sea urchins. The discovery of the regulation of kelp forest by sea otter predation on herbivorous urchins was strong support for the HSS hypothesis and for Paine's keystone species concept. [Figure 6.3]

In ecological terms, the predatory sea otters have a cascading effect on multiple trophic levels below them. Paine coined a new term to describe the strong, top-down effects that he and others had discovered when species were removed or reintroduced: he called these effects *trophic cascades*.

The discovery of trophic cascades was very exciting. The many indirect effects caused by the presence or absence of predators (starfish, sea otters) were surprising, because they revealed previously unsuspected—indeed, unimagined—connections among creatures. Who would have thought that the growth of kelp forests depended on the presence of sea otters? These dramatic and unexpected effects raised the possibility that, unbeknownst to biologists, trophic cascades were operating elsewhere to shape other kinds of communities. And if they were, then trophic cascades might be general features of ecosystems—rules of regulation that governed the numbers and kinds of creatures in a community.

Indeed, many trophic cascades have been discovered in all sorts of habitats. Just a few examples will suffice.

In a freshwater stream in Oklahoma, a predator-herbivore-algae trophic cascade regulates the abundance of minnows and plants. In pools in Brier Creek, Mary Power and colleagues noticed an inverse relationship between the number of bass and the number of minnows: the two animals co-occurred in just two of fourteen pools, and only then after a large flood. Moreover, pools that contained bass were green with filamentous algae, while those lacking bass but with minnows were barren. These distributions suggested the possibility that bass, like sea otters, kept the herbivorous minnow population in check, which in turn enabled algae to grow.

To test this, Power removed bass from a green pool and made a fence down the middle. She added the minnows to one side and left the other as a control. The minnows chomped the algae down to the barren state. She then added three largemouth bass to a minnow pool. Within just three hours, the minnows moved to the shallow

FIGURE 6.3 The influence of sea otters on sea urchins and kelp forests. (Top) In the presence of sea otters, sea urchin populations are controlled, which allows for kelp forests to grow. (Bottom) In the absence of sea otters, urchins proliferate, forming "barrens" that lack kelp.

Photos courtesy of Bob Steneck.

end of the pool where the bass could not get to them; within a few weeks, the pool was green. These results not only demonstrated the existence of a bass-minnow-algae cascade, but they also showed that predator avoidance could have a similar impact as predation itself.

Similar predator-herbivore-plant trophic cascades as the bass-minnow-algae and otter-urchin-kelp trios have been described on land. Isle Royale in Lake Superior, Michigan, was recolonized by moose and wolves in the first half of the twentieth century. Long-term studies have revealed that wolves positively influence fir tree growth by controlling the density of the moose, which feed extensively on the firs. And in Venezuela, trophic cascades were revealed when numerous predator-free islands were created in Lago Guri by the flooding of a tropical forest. For instance, the elimination of the normal army ant and armadillo predators of leaf-cutter ants unleashed a leaf-cutter population boom that decimated tree growth. By indirectly affecting tree growth, these predators also influence the habitat available for many other creatures.

The logic of these freshwater and terrestrial cascades looks like:

Predator	Bass	Wolves	Army Ants, Armadilloes
	⊥	⊥	⊥
Herbivore	Minnows	Moose	Leaf-cutter ants
	⊥	⊥	⊥
Producer	Algae	Fir trees	Trees

I have drawn these cascades in a way that emphasizes the top-down negative regulatory interactions between organisms at different trophic levels. But I must underscore that these are oversimplifications in two respects. First, most organisms are not in simple linear food chains, but as Elton knew well, are parts of food webs with more members and interactions. And second, all ecosystems are positively regulated to some degree from the bottom up. Without sunlight, there would be no plants; without plants, there would be no food for the herbivores; without herbivores, there would be no prey for the predators. That being said, the conceptual advance spurred by HHS and Paine was to flip traditional thinking upside down and to reveal strong indirect effects of predators on producers.

Trophic cascades are dynamic—not static—features of ecosystems. Indeed, trophic cascades have also been observed to flip, even without direct human intervention. By the 1970s, it was estimated that sea otter populations along the southeastern Alaskan coast had rebounded to about 100,000 animals. But the populations of sea otters has subsequently dropped dramatically from Castle Cape (south Alaska Peninsula) to Attu Island (Aleutian Island chain). While various possible culprits have been considered, Jim Estes and his colleagues have implicated killer whales as the main suspect. They suggest that the orcas recently turned to preying on sea otters when their preferred prey (sea lions and whales) became scarce. In so doing, the killer whales have turned a three-level cascade (left) into a four-level cascade (right), and the populations at lower levels have inverted—the coastal habitats are again carpeted with sea urchins and deforested of kelp.

$$
\begin{array}{ccc}
 & & \textbf{Killer whales} \\
 & & \perp \\
\textbf{Otters} & & \text{Otters} \\
\perp & \Rightarrow & \perp \\
\text{Sea urchins} & & \textbf{Sea urchins} \\
\perp & & \perp \\
\textbf{Kelp} & & \text{Kelp}
\end{array}
$$

NOT ALL ANIMALS ARE CREATED EQUAL

What began with the tossing of starfish has led to two fundamental insights about how nature works—two rules about the regulation of populations, our first two Serengeti Rules:

SERENGETI RULE 1

Keystones: Not all species are equal

Some species exert effects on the stability and diversity of their communities that are disproportionate to their numbers or biomass. The importance of keystone species is the magnitude of their influence, not their rung in the food chain.

It is important to stress that not all predators are keystones, and that not all keystones are predators. Nor do all ecosystems necessarily have keystones. But we will meet a couple more keystone species in Chapter 7.

SERENGETI RULE 2

Some species mediate strong indirect effects through trophic cascades

Some members of food webs have disproportionately strong (top-down) effects that ripple through communities and indirectly affect species at lower trophic levels.

Such cascades may exist wherever multiple pairs of strong regulatory interactions link trophic levels. These can be predator-predator, predator-herbivore, or herbivore-producer pairings.

It is important to emphasize that most species in a community do not exhibit strong interactions. In another colossal effort spanning several years, Paine investigated the strength of herbivore-producer interactions on Tatoosh Island. He found that most species interacted weakly or negligibly. Paine summed up that hard-earned knowledge in a quote from George Orwell's *Animal Farm*: "Some animals are more equal than others." I would add, some naturalists are also more equal than others. [Figure 6.4]

Just as the discovery of oncogenes and tumor suppressors revealed how not all genes are equal in the regulation of cell number, the discovery of keystone species and trophic cascades has revealed that in a community of organisms, not all animals are equal with respect to the regulation of populations. And just as focusing on those genes simplifies and defines the task for the oncologist, focusing on keystone species and trophic cascades simplifies the task for the ecologist in understanding the structure and regulation of an ecosystem.

So now, with new eyes, let's go back to where I started this book, and where our ancestors started—the magnificent Serengeti.

FIGURE 6.4 Robert Paine at Mukkaw Bay.

Photo courtesy of Kevin Schafer/Alamy.

CHAPTER 7
SERENGETI LOGIC

Africa is the continent most remarkable for the number
and variety of its large mammals. . . . Many have marveled
at this abundance of large life . . . but why is it so?

—JULIAN HUXLEY

The verandah outside Southern Highlands boarding school in Sao
Hill, Tanzania, was popular with visitors. At night, the lights drew
huge dung beetles, for which eight-year-old Tony Sinclair snuck out
of his dormitory to collect and keep as pets. One night, the young-
ster was stalking the bugs when he found himself face-to-face with a
leopard that was doing the same. The two hunters backed away from
one another very slowly.

Sinclair has been fascinated with animals for as long as he can
remember. It did not matter what sorts of creatures—beetles, birds,
chameleons—so long as they moved. Although born in Zambia (formerly Northern Rhodesia), and raised in Dar es Salaam (Tanzania,
formerly Tanganyika), his first eyeful of large African mammals in
the wild came on a stopover in Kenya when he was eleven years old.
Game parks were still a new concept in Africa in 1955, when Sinclair
visited a reserve outside Nairobi. He was awestruck by the great beasts.

After attending boarding school in England, Sinclair enrolled at
Oxford to study biology. On his second day on campus, he heard that

FIGURE 7.1 Migrating wildebeest, Serengeti National Park.

Photo courtesy of Anthony R. E. Sinclair.

there was a zoology professor who had students in the Serengeti. Sinclair had never been there and was eager to get back to Africa, so he went to see the professor, Arthur Cain. The don was taken by surprise.

"Oh well, I am going next year, better come with me," Cain said half-heartedly. But Sinclair came back every couple months to repeat his request. In the meantime, he discovered other opportunities at Oxford. He became good friends with Robert Elton, Charles Elton's son, and enjoyed family lunches with the famous ecologist at his home.

The next summer, on June 30, 1965, Sinclair crossed over the Mara River from Kenya into the Serengeti for the first time. His assignment was to be Cain's assistant, who had come to study the flocks of European and Asian birds that migrated through the park. But for the first three days, Sinclair and a companion drove all over 8,000 square miles of the Serengeti: across the plains; through the savanna and woodlands; and past massive herds of gazelles, zebras, and wilde-beest, dozing lions, and flocks of pink flamingos gathered on shim-mering lakes. The varied and beautiful landscape, the enormous vari-ety of plants and animals, and the great herds made the Serengeti, in Sinclair's eyes, the most extraordinary place on the planet. After just three days in this wonderland, Sinclair decided that he would spend the rest of his life studying the Serengeti to try to understand "why it was like the way it was."

The Serengeti has cast a spell on many who set their eyes upon it. The American hunter Stewart Edward White crossed into the North-ern Serengeti in August 1913, the first Anglo-explorer to enter the remote wilderness and to describe what he found:

> I set out by compass, bearing for a river called the Bologonja. . . .
> Went for miles over rolling burnt-out desert on which roamed
> a few kongoni and eland. Then saw the green trees of my river,
> walked two miles more—and found myself in a paradise.
>
> It is hard to do that country justice. From the river it rolls away
> in gentle, low-sloping hills as green as emeralds, beneath trees
> spaced in a park. One could see as far as the limits of the horizon,
> and yet everywhere were these trees, singly, in little open groves;
> and the grass was the greenest green.
>
> Never have I seen anything like that game. It covered every hill,
> standing in the openings, strolling in and out among the groves,
> feeding on the bottom lands, singly, or in little groups. It did not
> matter in which direction I looked, there it was; as abundant one
> place as another. Nor did it matter how far I went, over how many
> hills I walked, how many wide prospects I examined, it was always
> the same. . . . One day I counted 4,628 head! . . . I moved among
> those hordes of unsophisticated beasts as a lord of Eden would
> have moved.

White's thoughts on the discovery of such abundant wildlife quickly turned to their exploitation, which was the general mentality across colonial Africa at the time:

> And suddenly I realized again that in this beautiful, wide, popu-lous country, no sportsman's rifle has ever been fired. It is a virgin

game country, and I have been the last man who will ever discover one for the sportsmen of the world. There is no other available possibility for such a game field in Africa unexplored.

After World War I, Tanganyika Territory passed into British hands. In 1929, the government sent Julian Huxley, Charles Elton's former tutor, on a mission to East Africa to advise them on priorities and policies for their territories. After a four-month tour through Uganda, Kenya, and Tanganyika, the biologist thought that there were more valuable purposes for African wildlife than as quarry for sportsmen. In his 448-page account of his journey, *Africa View*, Huxley recommended that the Serengeti and other vast lands be preserved as national parks and game sanctuaries:

> In her large animals East Africa has a unique possession; if she allows them to be destroyed, they can never be replaced.... Humanity does not live by bread alone; in East African wilds a stream of men and women down the generations may find quickening, refreshment, inspiration.

Huxley was a schoolmate and friend of safari guide and hunter Denis Finch-Hatton, who was later made famous as Karen Blixen's lover in her memoir and the film *Out of Africa*. Finch-Hatton, whose clients included the future King Edward VIII, was a private man who eschewed the limelight. But he was appalled by the excessive killing in the Serengeti by tourist hunters. He wrote to the *Times* of London to condemn the "orgy of slaughter" and to push for protection of the Serengeti "before it was too late." Parliament took up the issue, and largely due to Finch-Hatton's efforts, the Serengeti was included in a Closed Reserve in 1930. In 1937, part of this area was declared a game sanctuary; in 1951, Serengeti National Park was established; and in 1981, it was declared a World Heritage site by the United Nations—a special status reserved for natural or cultural sites of outstanding universal value.

And indeed, the Serengeti is biologically very special. It is a vast ecosystem of almost 10,000 square miles that is bounded by natural barriers on all sides. It is home to one of the last remaining great concentrations of large mammals, so-called megafauna that have largely disappeared from or been exterminated on other continents. It is the site of one of the last remaining mass animal migrations on land. And, among all of its many mammals, it holds special significance for one—us. It is, as biologist Robin Reid put it, "the savanna of our birth," as it has been home to our ancestors for more than 3 million years. Its hippos, giraffes, elephants, and rhinos are descendants of the same sorts of creatures that ancient humans saw.

Once the park was established, along with the tourists came the biologists, who asked the obvious question: just how many creatures are there in the enormous Serengeti? In 1957, Bernard Grzimek, the director of the Frankfurt Zoo, and his son Michael were invited by the director of Tanganyika's National Parks to undertake the first detailed survey of Serengeti wildlife. For two weeks in January 1958, they flew their zebra-striped Dornier 27 very slowly, 150–300 feet above the vast plains, counting everything on four legs that they could see. With Teutonic precision, they reported 99,481 wildebeest, 57,199 zebra, 194,654 Thomson's and Grant's gazelles, 5,172 topi, 1,717 impala, 1,813 black buffalo, 837 giraffes, and 60 elephants. Altogether, they calculated 366,980 large mammals living in the park itself, but allowed that they might have missed perhaps 10,000 animals. They noted many thousands more roaming outside the park boundaries.

The numbers seemed very large, "almost inconceivable" to the Grzimeks. "Were there enough plains, mountains, river valleys and bush areas to maintain the last giant herds still in existence?" they worried. It is a question that each successive generation of Serengeti scientists has been asking, and fretting about, ever since the Grzimek's survey.

Ironically, the great masses of wildlife that so enthralled White, Finch-Hatton, Huxley, the Grzimeks, and Sinclair represented the Serengeti at only a fraction of its full glory. When Sinclair arrived, major changes were beginning to be detected among the big animals. A survey in 1965 counted about 37,000 buffalo in the system compared to just 16,000 four years earlier. Some of the scientists working in the Serengeti suggested to Sinclair that perhaps he might want to study the rapidly expanding buffalo population as a PhD project. "Can a bird man do that?" they teased.

Yes he could, he assured them. Sinclair was not stuck on any group: all animals interested him. He would turn from birds to buffalo, which would in turn give him the key clue to discovering why the Serengeti was the way it was and why it was changing—the Serengeti Rules that regulated not only buffaloes but also all sorts of herbivores and carnivores, and even the trees.

WHY DO MORE BUFFALO ROAM?

Numbers. That's all Sinclair had to go on when he started his work in earnest in October 1966, so little was known about the ecology of any of the wildlife in the Serengeti (George Schaller started his pioneering studies of Serengeti lions the same year). But those numbers of animals posed great mysteries. Why was there a certain number of animals at a given time and place? And what explained the great differences in numbers among different kinds of animals? Why were there so many wildebeest, for example, and so few hartebeest, its close relative?

Before he could tackle such broad questions, he needed to know that the trend for buffalo was real, not a miscount or a short-term aberration, and to learn a lot more about how buffalo lived and died.

Sinclair joined the buffalo census-taking effort in 1966 and led it from 1967 on. To count the buffalo, he first had to find them. Different species prefer different habitats. The three main habitats in the Serengeti where most of the animals live are distinguished by their vegetation, which is important, because that defines the different foods available to the grazers and browsers, and their predators. There are the grasslands, which are just that, vast nearly treeless plains covered with grass; there is the savannah, which is a grassland dotted with trees that are sparse enough to allow the grass to grow beneath them; and there is the woodlands, which are parts of the savanna where the trees are more dense. The buffalo preferred the open woodlands and avoided the treeless plains.

To count them, Sinclair and other spotters flew low over 4,000 square miles of woodlands in a grid pattern for three or four days, usually in the mornings while the animals grazed in the open, and photographed the herds. Overlapping photographs were then plotted

. onto maps of the Serengeti. Sinclair repeated the survey almost every year through 1972. The buffalo population was increasing. In fact, by 1969, there were so many buffalo (almost 54,000) that counting them all was just too laborious, so Sinclair switched to counting a sample in the northern woodlands and then extrapolating to the whole of the Serengeti. By 1972, he estimated the population at more than 58,000.

The steepest increase had occurred from 1961 to 1965, and the upward trend had continued over the next seven years. The next question was why—why were the buffalo increasing? The possible explanations that could account for the overall trend were an increase in fertility, or a decrease in death rate, or a combination of both. To distinguish among these possibilities, Sinclair first examined the fertility rate of female buffalo, but he found that it had remained stable over time.

He then examined the death rates among buffalo. Thousands of buffalo died in the Serengeti each year. Sinclair learned that he could determine buffaloes' ages by examining their teeth: younger animals' teeth emerged in an ordered sequence, and older animals' roots had alternating dark and light bands that marked the years (like tree rings). Sinclair examined the skulls of almost 600 buffalo that died on the Serengeti, and found that the highest death rates occurred in the first year of life and in animals older than fourteen years. By plotting the buffalo death rates against census records back to 1958, he figured out that juvenile mortality was much higher in 1959–1961 than in 1965–1972.

So there was the mystery: Why were more juvenile buffalo dying in those earlier years? And why did the juvenile mortality rate later drop?

There are three main ways for buffalo to die—from attack by predators, from disease, or from lack of food. Field observations suggested that predation by lions or hyenas was not the major factor in buffalo mortality. Nor could malnutrition explain the higher juvenile mortality in earlier years, as later years proved that the Serengeti could support many more buffalo. So that left disease. Buffalo, like most animals, are susceptible to a plethora of infectious diseases, but one suspect in particular stood out to Sinclair.

Known as rinderpest or "cattle plague," the often-fatal disease is caused by the rinderpest virus, a sister to the human measles virus.

The disease was known for centuries in Asia and India. In 1889, the virus arrived in East Africa. It is believed that it was first introduced when Italian soldiers brought infected cattle from India or Arabia to Ethiopia during a military campaign. It then made its way to the Serengeti via Maasai cattle, and it promptly devastated the wild ruminants. In August 1891, German Oscar Baumann crossed the Serengeti and estimated that nearly 95 percent of the cattle, buffalo, and wildebeest had died. Periodic outbreaks in Serengeti animals were recorded over the next seventy years: during World War I, from 1929 to 1931, in 1933, in 1945, in 1957 and every year thereafter through 1961, with a severe outbreak in October 1960.

Sinclair wondered: Had rinderpest kept the buffalo population suppressed in earlier years? And did its recent absence allow their rapid increase? To find out, Sinclair looked for traces of rinderpest infection in buffaloes of different ages. Animals that are exposed to the virus make antibodies against it in their blood serum, and these antibodies are easily measured in the laboratory. If his idea was right, older animals should have antibodies and younger animals should not.

Sinclair knew that virologist Walter Plowright, who had developed a new vaccine against the virus, had been monitoring rinderpest infection in East Africa for many years. Sinclair had given Plowright buffalo serum samples that he had taken in the late 1960s. He went to see Plowright at the East African Veterinary Research Organization lab in Muguga, outside Nairobi. There he learned that, to his good fortune, the skulls had been kept of the animals from which serum samples had been taken. Sinclair could therefore date the buffalo and test his rinderpest theory.

He found that while most of the animals born in 1963 or earlier had antibodies against rinderpest, none of the buffalo born in 1964 or later had been exposed. Perfect! It was Sinclair's first eureka moment.

The correlation between the presence or absence of rinderpest and the repression or expansion, respectively, of buffalo populations immediately raised another possibility—that the same explanation would apply to wildebeest populations. The wildebeest population had more than tripled since 1961. Sinclair examined antibody data for wildebeest and also found a sharp cutoff in the presence of

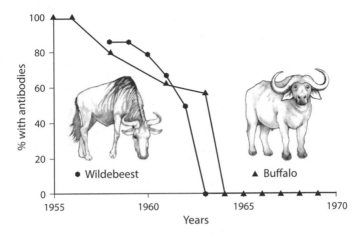

FIGURE 7.2 The elimination of rinderpest virus from Serengeti wildebeest and buffalo. Circulating antibodies to the virus disappeared from Serengeti wildebeest and buffalo populations in 1963 and 1964, respectively, indicating that the disease had been eliminated from the Park.

Figure drawn by Leanne Olds based on data in Sinclair (1979).

antibodies in wildebeest, with no animals born in 1963 or later showing evidence of exposure to the virus. [Figure 7.2] Moreover, Sinclair realized that the impact of the virus was species-specific. The population of zebras, which are not ruminants and are not susceptible to rinderpest, had remained steady over the decade.

Sinclair's evidence for the disappearance of rinderpest in buffalo and wildebeest overturned the prevailing opinion as to the origin of new rinderpest infections in East Africa. It was thought that wildlife was the source of outbreaks of cattle plague. The ongoing vaccination program in East Africa targeted only domesticated cattle, but it had the added effect of eliminating the virus from wild animals. This proved that cattle were the reservoir for rinderpest, not the wildlife.

Sinclair had solved the mystery of the rapid expansion of the ruminant buffalo and wildebeest. The rinderpest virus was acting as a microscopic keystone in the community. Its presence negatively regulated the ruminants; its suppression allowed them to erupt:

Cattle
Rinderpest   (vaccination)   Rinderpest
⊥   ⊥   →   ⊥   ⊥
Buffalo Wildebeest   **Buffalo Wildebeest**

The enormous impact of rinderpest shows that it is not only predators that can act as keystones, pathogens can also have a disproportionate influence on communities. And just as for predators, their introduction or their elimination can have cascading effects through ecosystems. Rinderpest had been repressing the Serengeti for seventy years. As Sinclair would also discover, the release of the ruminants unleashed a surprising array of changes.

130,000 TONS OF WILDEBEEST

In a perverse way, rinderpest was a gift to the ecologist. Like the removal of starfish or the reintroduction of sea otters, the virus was the perturbation (albeit an accidental one) that allowed Sinclair and other scientists to see how the Serengeti community worked. By 1973, the population of wildebeest had reached an astonishing 770,000 animals, but unlike the buffalo, they had not yet shown signs of leveling off. The wildebeest were eating most of the food, and themselves were important items on carnivores' menus. It dawned on Sinclair that if he wanted to understand the Serengeti, he needed to pay more attention to the wildebeest.

But working in the Serengeti became more difficult in the mid-1970s—not because of the ruminants, but on account of *Homo sapiens*. In the late 1960s, Tanzania adopted an extreme form of socialism by collectivizing its agriculture, nationalizing banks and other corporations, and banning private forms of ownership. After years of tension with capitalist Kenya, Tanzania abruptly closed its borders in February 1977. With tensions raised and travel restricted, tourism in the Serengeti plummeted by more than 80 percent. The Mara section of the Serengeti ecosystem was in Kenya, so it was uncertain whether scientists could cross the border to take the animal censuses.

By 1977, the wildebeest had not been surveyed for four years. Sinclair and colleague Mike Norton-Griffiths, who was also a pilot, aimed to take a full count of the population. On the picture-perfect day of May 22, they took off from the airstrip near the Serengeti Research Institute and began to fly back and forth, working their way from north to south across the Serengeti. In addition to the vast herds of wildebeest, Sinclair noticed lines of trucks moving north toward the Kenyan border.

As they landed back at the airstrip, they were met by Tanzanian soldiers with guns drawn. An officer asked Sinclair and Norton-Griffiths what they were doing flying back and forth over their vehicles. They cheerfully replied that they were counting wildebeest. The officer did not believe them, asking how could they count animals from such a height? Sinclair replied that they were photographing them and would count them later. This did not go over well at all, especially when the officer learned that Norton-Griffiths had flown in from Kenya.

"So, you are from Kenya and you are photographing our army. I am arresting you for spying for Kenya," the officer said. Their plane was impounded, but Sinclair did manage to sneak the exposed films out of the camera.

The scientists were then confined to their house under guard except, they noticed, when the shift changed. After three days of captivity, the prisoners made their move. During a shift change, Sinclair and Norton-Griffiths dashed for the plane, hopped in and took off. Realizing they did not have enough fuel to make it back to Kenya, they decided to detour to Mary Leakey's camp at Olduvai Gorge in the hope that the paleontologist might have some fuel to spare.

She had some fuel, but she also had quite a story of her own to share. The previous year, at a site nearby called Laetoli, some of her team had stumbled upon some animal tracks that had been preserved in an ancient ashfall. Among the many tracks, they had found some remarkable, familiar-looking footprints. Leakey stunned Sinclair and Norton-Griffiths by showing them the first casts of what would turn out to be an eighty-eight-foot-long track of at least two sets of human-like footprints made 3.6 million years ago. The discovery of these ancient footprints would remove all doubt about the bipedal posture of our early ancestors.

A few weeks later, Sinclair and Norton-Griffiths got their own surprise when they developed the film from their census and discovered that the wildebeest population had reached 1.4 million animals—almost double what it had been four years earlier, and more than five times what it was in 1961. It was now the largest wild ungulate herd in the world. Other scientists had noticed a broad spectrum of other changes in the Serengeti during this period. For instance, the numbers of lions and hyenas increased. That certainly made sense, as there was more prey available. Those extra million or so wildebeest constituted about 130,000 tons of extra biomass—that could feed a lot of carnivores. But there were other, more puzzling changes, the causes of which when considered individually were not so obvious. For example, the number of giraffes also increased. Was there some connection between the giraffes and the other changes taking place in the Serengeti?

Indeed there was. The critical piece of the puzzle came from Mike Norton-Griffiths's studies, which revealed that the frequency and intensity of fires in the dry season had dropped dramatically in the Serengeti since 1963. Fire represses the regeneration of young seedlings. The reduction in fires meant that more young trees could grow, providing more food for the giraffes.

But why were the fires decreasing? Norton-Griffiths and Sinclair realized that they were staring at the explanation in their census data. The spike in wildebeest and buffalo populations meant that much more grass was being consumed by the herbivores, resulting in much less dry-season fuel. All the changes in the Serengeti were connected, they were all responses to the same perturbation: the elimination of the rinderpest virus, which had unleashed trophic cascades that affected herbivore, carnivore, and tree populations.

Take a few moments to follow the cascades and logic depicted in Figure 7.3. The most important revelation of these long-term studies is that, contrary to essentially every depiction on television, the real drama of the Serengeti is not a cheetah or lion chasing down a gazelle—it is a wildebeest munching on grass. For that mundane activity, multiplied by 1 million or more, triggers a cascade of interactions on the savanna that leads to more predators, more trees, more giraffes, and other species as well.

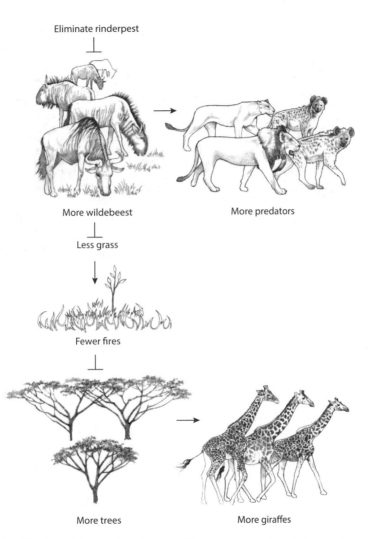

Eliminate rinderpest

More wildebeest

More predators

Less grass

Fewer fires

More trees

More giraffes

FIGURE 7.3 The real drama of the Serengeti. The elimination of the rinderpest virus and increased consumption of grass by the exploding wildebeest population unleashed trophic cascades that increased the populations of predators, trees, giraffes, and other species.

Illustration by Leanne Olds.

FIGURE 7.4 The explosion of trees on the Serengeti. Reduced fires led to increases in tree density as monitored by photographs over a twenty-one-year period.

Photos courtesy of Anthony R. E. Sinclair.

Of all the changes triggered by the wildebeest boom, the "outbreak of trees" most surprised Sinclair. Like the connection between sea otters and kelp forests, the connection between rinderpest and trees involved multiple levels of negative regulation—in this case triple-negative logic, that were not obvious. Indeed, for decades, researchers and conservationists had been fretting over the disappearance of mature trees on the Serengeti and blamed elephants for their loss. The possibility that more young trees were regenerating had largely been overlooked.

But Sinclair was not satisfied with a good correlation. To test whether tree populations were in fact expanding on the savanna, he set up a series of camera points from which he documented the changes in tree populations. "It only took a decade," he told me, to confirm that there was an "explosion" of several species of trees occurring across the Serengeti. [Figure 7.4]

On the treeless plains, the wildebeest have other important effects on plants besides fire suppression. Before their increase in numbers, the grass on the eastern plains grew to be fifty to seventy centimeters high. After their eruption, the grass grew to just ten centimeters. The shorter grass allows more light and nutrients for other plants, such that many more species of herbs grow. These herbs in turn support larger numbers and a more diverse community of butterflies.

Amazingly, the effects of the wildebeest and other grazers on the grasses are not entirely negative. Ecologist Sam McNaughton discovered that the major Serengeti grasses have adapted to the intense grazing by evolving a compensatory growth response that regenerates their above-ground parts. The grasses actually produce more food and are more abundant when grazed than when protected from grazing. In this fashion, the wildebeest positively regulate the formation of a dense "grazing lawn" that sustains them year after year (denoted by the symbol "↑◡" in the schematic below).

The wildebeest compete with other animals that also feed on the grass, such as grasshoppers. Both the number and diversity of grasshoppers decreased dramatically after the wildebeest boom, from more than forty species initially to about a dozen. Competition with wildebeest for food also appears to explain a reduction in the population of Thomson gazelles. In the four years during which the wildebeest population doubled, Sinclair and Norton-Griffiths found that the gazelle population had been reduced by half, from 600,000 to 300,000 animals. In contrast, the removal of buffalo from certain areas has revealed that they do not have such strong effects on other species.

Just like the mussels on rocky shores, the wildebeest are strong competitors for resources on the plains (denoted below with a double-headed arrow ↔), and their activity regulates the populations of species on the plains as well as the savanna:

Gazelles ↔ **Wildebeest** ↔ Grasshoppers

⊥

Grasses ↔ **Herbs** → **Butterflies**

↑◡

Competition is another major means by which the number and diversity of populations are regulated, and thus constitutes another Serengeti rule:

SERENGETI RULE 3

Competition: Some species compete for common resources

Species that compete for space, food, or habitat can regulate the abundance of other species.

The wildebeests' many direct and indirect effects on grasses, fire, trees, predators, giraffes, herbs, insects, and other grazers reveal that they are a keystone species in the Serengeti, with disproportionate impacts on the structure and regulation of communities. As Tony Sinclair put it, "Without the wildebeest, there would be no Serengeti."

But, you may now be wondering, what regulates the number of wildebeest? The population could not and did not go on expanding forever. In fact, 1977 marked their peak. What then, in the absence of rinderpest, curbed the wildebeest boom? And what about other species, like impala, buffalo, or elephants? What regulates their numbers?

The pursuit of these questions will lead us to another set of Serengeti Rules that regulate the numbers of many different kinds of animals, not just in East Africa, but across the world.

SIZE MATTERS: WHO IS ON THE MENU, AND WHO IS TOO BIG TO EAT?

To eat or to be eaten, that is animal life in a nutshell. In the absence of epidemics like rinderpest, this truth frames two fundamental ways in which specific animal populations could be regulated: by what there is to eat—the availability of food (from the bottom up in trophic terms); by being eaten by predators (from the top down); or some combination of both. For any species, the simple question is: which of the two is more important?

For most species in nature, the question is much easier to ask than to answer. Long-term observations are required, and preferably, experiments might be performed. Sinclair and his colleagues Simon Mduma and Justin Brashares examined forty years of data on the causes of death among Serengeti mammals. They found a striking correlation between adult body size and animals' vulnerability to predators.

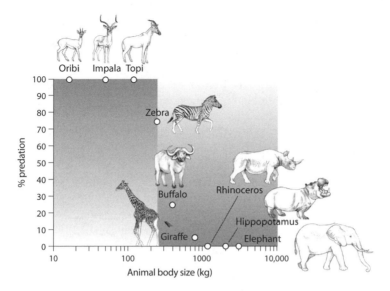

FIGURE 7.5 Predation rate depends on prey body size. Smaller antelope such as oribi, impala, and topi largely die from and are regulated by predation. Larger mammals such as giraffes, hippos, and elephants experience little or no predation; their populations are regulated by food supply.

Illustration based on data in Sinclair et al. (2010), drawn by Leanne Olds.

There was a fairly sharp threshold around 150 kilograms of body weight, below which smaller animals were generally regulated by predation, and above which larger animals were not. For example, most smaller antelopes, such as oribi (eighteen kilograms), impala (fifty kilograms), and topi (120 kilograms), die from predation (see Figure 7.5, top left). Generally speaking, the smaller the animal, the more predators it has. For example, of the ten mammalian carnivores in the Serengeti (including wild cats, jackals, cheetahs, leopards, hyenas, and lions), oribi are preyed on by at least six, as well as by eagles and pythons. [Figure 7.5]

But larger mammals, such as buffalo, experience much less predation (by lions only) and adult giraffe, rhino, hippos and elephants essentially none at all (see Figure 7.5, bottom right). The latter herbivores, so-called megaherbivores, appear to have escaped regulation by predators by evolving large bodies (and defenses) that make them just too difficult or dangerous to bring down, even by lions. Since elephants and the other large mammals above the size limit are not regulated from the top down by predators, it follows then that they must be regulated from the bottom up, by the availability of food.

The body size threshold is an interesting correlation, but was there any way to test it in Paine-like fashion—to "kick" the Serengeti and see what happened? There was. Unfortunately, the kick came in the form of increased poaching and poisoning that eliminated the majority of lions, hyenas and jackals from the northern Serengeti from 1980 to 1987. Sinclair and his colleagues compared the prey populations before and after the carnivores were reduced, and again once the predators returned in later years. All five small-bodied prey they monitored (oribi, Thomson's gazelle, warthog, topi, and impala) increased in numbers during the period of predator removal, but the giraffe population did not. And all five small-bodied populations decreased again once the predators returned, demonstrating that these species, but not the giraffe, are negatively regulated from the top down by predators.

These observations about Serengeti predators and their prey provide quantitative and experimental support for an inference Elton made almost eighty years earlier (without the benefit of having seen anything like the Serengeti): "the size of the prey of carnivorous animals is limited in the upward direction by its strength and ability to catch the prey, and in the downward direction by the feasibility of getting enough of the smaller food to satisfy its needs." They reveal a specific rule about how body size can determine whether a population may be subject to regulation by predation:

SERENGETI RULE 4
Body size affects the mode of regulation

Animal body size is an important determinant of the mechanism of population regulation in food webs, with smaller animals regulated by predators (top-down regulation) and larger animals by food supply (bottom-up regulation).

Now, if being too large to be killed is such a decided advantage, one might think that all species would have evolved in that direction in such a predator-rich habitat. But they didn't. Nor is the Serengeti carpeted with elephants or buffalo. Their numbers are regulated, too, but how does the regulation of such large animals work? It turns out that, although we are seeking to explain regulation on the largest scale in nature, the mechanism involved is already familiar from molecular biology.

FEEDBACK REGULATION OF THE BEASTS

Sinclair's surveys showed that after a dramatic surge following the disappearance of rinderpest, the buffalo population leveled off in the 1970s. The story of Serengeti's elephants is also one of rebound, but from a different plague. The ivory trade in the nineteenth century decimated the population such that the animals were scarce in the first part of the twentieth century. The Grzimeks counted just sixty elephants in the southern part of the park in 1958, but the population throughout the Serengeti system expanded from the early 1960s to mid-1970s to several thousand elephants and held relatively steady for years.

When Sinclair plotted the rate of increase of each species against population size, he obtained similar-looking graphs. [Figure 7.6] What the curves revealed was that the *rate* of increase of each species was higher when their numbers were fewer, decreased as the populations grew, and then turned negative (the populations decreased). In other words, the rate of change in the population depended on its density.

This phenomenon is known as *density-dependent regulation*. It has been appreciated since the writings of social economist Thomas Malthus that populations will increase indefinitely unless something prevents them from doing so. Imagine, however, a group of large animals in a fixed space, like goats in a pasture. If the population starts out small in number, it can expand as rapidly as the animals can reproduce. But as the number of animals increases, space or food begins to run low. If the population has overshot the capacity of the habitat, it will contract; eventually it will level out to the maximum that can be supported on finite resources.

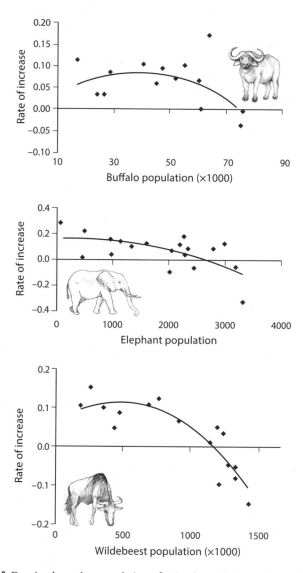

FIGURE 7.6 Density-dependent regulation of animal populations. As populations of buffalo, elephants, and wildebeest increased in the Serengeti, their rate of increase slowed and then turned negative (their numbers shrank).

Illustration based on data in Sinclair et al. (2010), Sinclair (2003), and Sinclair and Krebs (2002), and drawn by Leanne Olds.

Density-dependent regulation is a form of negative feedback regulation. Just as the buildup of the products of enzyme reactions can feedback to inhibit the process that produces them, the buildup of animal populations can slow or even reverse the processes that would produce more animals. Sinclair investigated how this negative feedback regulation worked in buffalo by examining fertility and mortality rates. He discovered that a greater *proportion* (not just the number) of adults died of malnutrition as the population increased.

Sinclair, Simon Mduma, and their colleague Ray Hilborn found that the same density-dependent mechanism curbed the population of migrating wildebeest. As the population approached 1 million, the rate of increase slowed and then turned negative (see Figure 7.6, bottom). To figure out what caused this density-dependent regulation, they examined forty years of records of the wildebeest census and the animals' causes of death. They discovered that while predation was significant (25–30 percent of deaths), most wildebeest died of poor nutrition when the population was large. By scrutinizing the records of rainfall and grass biomass on the Serengeti, they found that this malnutrition correlated with the amount of food available per capita in the dry season.

While the Serengeti is an immense, productive land, the dry season is a critical period when quality forage is less abundant, and animals are more vulnerable. This vulnerability was underscored when a natural "experiment" began to unfold in 1993—the worst drought in thirty-five years struck the Serengeti. Food supplies during the longer dry season dropped to a fraction of typical years. Sinclair, Mduma, and Hilborn were eyewitnesses to the mass starvation as up to 3,000 wildebeest were dying every day in November; about 30 percent of the wildebeest succumbed during the dry season, and the population dropped to under 1 million animals.

A tragic episode, but there is an important flip-side to density-dependent regulation to appreciate. Once the population dropped, there was more food available per capita in subsequent seasons, and the population stabilized. Density-dependent regulation has the "virtue" of buffering change in both directions, slowing expansion as populations get large, and slowing declines as populations drop. It has been likened to a thermostat that triggers cooling when temperature exceeds a set point and heating when it drops below.

Food is not the only factor that can regulate in a density-dependent way; predators can also keep a population from increasing, but as the population drops and prey becomes more scarce, predators may turn to alternative, more abundant prey, which allows the primary prey population to rebound (and not to go extinct). Competition for space, such as nesting sites or territory among predators, can also regulate populations in a density-dependent manner. Feedback regulation through density-dependent factors is a widespread mechanism that regulates animal numbers:

SERENGETI RULE 5

Density: The regulation of some species depends on their density

Some animal populations are regulated by density-dependent factors that tend to stabilize population size.

We have seen two major ways in which animal numbers are regulated—by predators and by the availability of food. And we have seen one way that animals escape predation—by evolving larger body size. Is there some way that they can also circumvent, at least to some degree, the limitations of food supply?

Actually, there is a way to do both—to elude predators and to access more food—and it explains the greatest spectacle on the Serengeti.

MIGRATION: HOW TO EAT MORE WITHOUT GETTING EATEN

Back to some numbers you know well by now: 60,000 buffalo, over 1 million wildebeest. The 450-kilogram buffalo are much less vulnerable to predators than the 170-kilogram wildebeest, but there are a heck of a lot more wildebeest on the Serengeti than buffaloes. Besides body size, what else distinguishes these two species?

One stays put, the other doesn't.

Could migration account for the vast difference in numbers between the most abundant sedentary and migratory species on the Serengeti?

Since two major ways to regulate population size are food limitation and predation, we need to know the effect of migration on each mode of regulation. That is just what Sinclair and colleagues have done. The dietary advantages of migration are pretty straightforward. The wildebeest follow the rains in a 600-mile-long annual circuit around the Serengeti, moving into the green, highly nutritious, short-grass plains in the wet season. That is a short-lived resource that nourishes their developing calves and is not exploited by sedentary species. Then, as the plains dry out, they move to the tall-grass savanna and woodlands, which receive more rainfall than the open plains.

The predation side of the equation requires a bit more exploration. Wildebeest are prey for lions and hyenas. But in discussing the body sizes of prey earlier, I deliberately omitted the statistics for wildebeest. The reason I did so is that the figure depends on the population of wildebeest. As it so happens, there are two kinds of wildebeest on the Serengeti: the vast migratory herds, and smaller pockets of "resident" populations that remain year-round in certain parts of the system (near stable sources of water). Predation on those resident populations accounts for almost 87 percent of their deaths, whereas predation on the migrants accounts for only about one-fourth of migrant deaths. Moreover, only about 1 percent of the migrants are taken in a given year, while up to 10 percent of residents may be killed. The migrants, therefore, experience much less predation per capita. Studies of lion and hyena behavior explain why they cannot take advantage of all that meat on the go: the predators cannot follow the herds, because they are confined to territories to raise and protect their own young.

The combined effect of evading predators and accessing more food allows the migratory wildebeest to achieve a much greater density (about sixty-four animals per square kilometer) than that of the resident populations (about fifteen animals per square kilometer). The large numbers of two other Serengeti migrants—zebras (200,000) and Thomson gazelles (400,000)—relative to all other sedentary species is also consistent with migration posing a major advantage. Elsewhere in Africa, such migrants as the tiang (a race of topi) and white-eared kob of Sudan also outnumber the most abundant sedentary species by at least a factor of ten.

Migration, then, is another ecological rule, or more aptly a rule-breaker, a way of exceeding the limits imposed by density-dependent regulation:

SERENGETI RULE 6

Migration increases animal numbers

Migration increases animal numbers by increasing access to food (reducing bottom-up regulation) and decreasing suscep-tibility to predation (reducing top-down regulation).

DIFFERENT RULES, SIMILAR LOGIC

Fifty years after Sinclair [Figure 7.7] first stepped on the Serengeti, he is still here. After studying its migratory inhabitants for so long, he

FIGURE 7.7 Tony Sinclair on the Serengeti.

Photo by Anne Sinclair, courtesy of Anthony R. E. Sinclair.

has become one. He and his wife Anne built a house on the shore of Lake Victoria at the western edge of the Serengeti, to which they return every year.

Thanks in large part to "Mr. Serengeti"—a nickname used with respect and affection by his colleagues—we now understand the rules of this extraordinary place. We now know about the food webs, keystone species, trophic cascades, competition, density-dependent regulation, and migrations that determine why it is the way it is: why there are so many zebras, and not so many elephants; why predators control the impala and topi, but not the giraffe or hippos; why there are more trees and butterflies today than fifty years ago, but fewer grasshoppers; and why the long-faced, uncharismatic wildebeest and their movements are, in Sinclair's words, the Serengeti's "lifeblood."

But these Serengeti Rules are not just for Serengeti games; they are general rules that apply to ecosystems everywhere. Moreover, when we compare them to the general rules of regulation and the logic of life at the molecular scale (shown Chapter 3), they turn out to be remarkably similar. The specific means of regulation are different in ecology (predation, trophic cascades, etc.), but positive and negative regulation, double-negative logic, and feedback regulation control numbers at both scales:

GENERAL RULES OF REGULATION AND THE SERENGETI RULES

Positive regulation

A → B   Bottom-up regulation of higher trophic levels

Negative regulation

A ⊣ B   Top-down regulation by predators; competition

Double-negative logic

A ⊣ B ⊣ C   Trophic cascades: A has strong indirect effect on C by regulating B

Feedback regulation

A → → → A   Density-dependent regulation; the growth rate declines as population increases

And like the specific molecular rules that govern our health, these ecological rules are also rules to live by. For as we shall see next, when they are broken, bad things happen in *our* world. And just as with molecular rules, understanding these rules of ecological regulation enable us to diagnose what is ailing ecosystems, and potentially, to cure them.

CHAPTER 8
ANOTHER KIND OF CANCER

It is failures in regulation of numbers of animals which
form by far the biggest part of present-day economic
problems in the field.

—CHARLES ELTON

At 1:20 a.m. on Saturday, August 1, 2014, the city of Toledo, Ohio, issued an urgent alert to all residents:

DO NOT DRINK THE WATER
DO NOT BOIL THE WATER

Chemists at the city's water treatment plant had detected dangerous levels of a nasty toxin in the water supply, a toxin that could not be destroyed by boiling, but instead would only become more concentrated.

The metropolitan area of one-half million people was brought to a standstill. Restaurants, public buildings, and even the city zoo closed. People quickly bought up whatever bottled water was on store shelves. The governor of Ohio declared a state of emergency. The National Guard was enlisted to truck in water and portable water treatment plants. The national and international news media covered the story of a modern American city without the 80 million gallons of water it needed daily. It was not the sort of attention the long-struggling, rust-belt city wanted.

I was paying extra attention. It is a city, and water, that I know very well. I was born and raised in Toledo, which is situated on the southwest shore of massive Lake Erie. My friend Tom Sandy and I often went snake hunting near the edge of the lake, and the thrill of those catches played a large part in my desire to become a biologist. But never in my entire childhood did I dip one toe in the lake's waters. Nor would I eat anything anyone caught in it.

The lake was notorious for its level of pollution while I was growing up in the 1960s and early 1970s, so notorious that Dr. Seuss singled it out in his environmental fable *The Lorax* (1971):

> You're glumping the pond where the Humming-fish hummed!
> No more can they hum, for their gills are all gummed.
> So I am sending them off. Oh, their future is dreary.
> They'll walk on their fins and get woefully weary
> In search of some water that isn't so smeary.
> I hear things are just as bad up in Lake Erie.

Spurred by the dire condition of Erie and other lakes, the US Congress passed the 1972 Clean Water Act that authorized the Environmental Protection Agency to regulate the discharge of pollutants into waterways and to set the acceptable limits for water quality for humans and aquatic life. In 1972, the United States and Canada also signed the Great Lakes Water Agreement, which promoted a coordinated effort to reduce the loads of chemicals that were being dumped and washed into the Great Lakes.

Algal populations dropped, fish populations grew. The recovery of Lake Erie was so dramatic that in 1986, Dr. Seuss even agreed to remove its mention from later editions of *The Lorax*.

But Lake Erie is again getting glumped. The immediate culprit is a tiny, single-celled, blue-green algae called *Microcystis* that forms thick mats that can cover many miles of lake surface. In 2011, the lake experienced its largest bloom in history, a green carpet up to four inches thick stretched for 120 miles along the southern shore, from Toledo to Cleveland. In 2014, the thick pea soup formed right on top of the Toledo water treatment plant's main intake pipe. [Figure 8.1]

The blooms contain astronomical numbers of algae. Under typical conditions, there may be just a few hundred algal cells in a liter

FIGURE 8.1 Lake Erie algal bloom near Toledo, Ohio, August 2014.

NASA satellite photograph taken on August 1, 2014.

of lake water. In a bloom, this can rocket to more than 100 million cells per liter. The 2011 bloom may have contained as many as 1 quadrillion (1 million trillion) to 1 quintillion (1 billion trillion) toxin-producing cells in all.

Like a tumor metastasizing through the human body, the mass of algae sows destruction as it spreads through the body of the lake. The massive overgrowth of algae is indeed an ecological cancer.

When cancer spreads in a person, it can invade and cripple the organs that maintain the body's homeostasis. When it hits the bone marrow or lungs, the body can starve for oxygen; when it invades the digestive organs, the body is starved for nutrients; and when it infiltrates the liver and bone, it can throw off the delicate balance of key chemicals in the bloodstream. Similarly, the algal mass kills by blocking vital functions in the lake. The toxin(s) it produces is highly toxic to fish and other wildlife, wreaking havoc on the food chain. And as the algae die off, they sink to the lake bottom, where bacteria that decompose them use up the lake's supply of oxygen—suffocating fish and other creatures, and creating an uninhabitable dead zone with altered water chemistry.

Lake Erie is not the only large body of water that is in critical condition. It has plenty of company, including Lake Winnipeg in Canada, Lake Taihu in China, and Lake Nieuwe Meer in the Netherlands. Nor are these the only ecosystems suffering from the overgrowth of some creature. Cancer takes various forms in different parts of the biosphere. I will look at a few more cases before asking what sorts of rules have been broken that can make lakes, fields, bays, and the savannah sick, and in Chapters 9 and 10 I will show how that knowledge can be used to heal them.

PESTILENCE

Fly over or visit any of the sixteen countries of tropical Asia, and it is clear what the people are eating. From India to Indonesia, mile after mile of rice fields sprawl across valleys and up and down terraced hillsides. In Cambodia, for example, rice production alone occupies over 90 percent of the total agricultural area. The grain is a critical staple now for almost half of humanity. Over 30 percent of all calories consumed in Asia come from rice, and in some countries, such as Bangladesh, Vietnam, and Cambodia, the grain provides over 60 percent of daily intake.

Rice has been cultivated in Asia for more than 6,000 years, but today's lush fields are the products of the Green Revolution of the 1960s. Facing the possibility of massive famine due to drought, crop failures, and a booming population, new genetically improved rice varieties were developed, and more productive farming methods were introduced, including the routine application of fertilizers and pesticides. Within ten years, more than a quarter of all farms were using new rice strains, and many farmers across Asia saw their rice yields per acre nearly double.

But in the mid-1970s, many bright green paddies in the Philippines, India, Sri Lanka, and elsewhere across tropical Asia were turning orange-yellow, then brown. In 1976, disaster struck Indonesia. More than 1 million acres of crops were afflicted. In a region where farmers rely on their crops to feed their families for the year, or for most of their annual income, the situation was dire.

The culprit was a tiny insect called the brown planthopper. Although just a few millimeters long, individual females that land on a plant may lay up to several hundred eggs that then hatch into hungry nymphs that feed on the growing rice plants. [Figure 8.2] The little bugs suck the sap; then the plants' leaves turn yellow, dry out, and die, producing a characteristic "hopperburn." With a rapid generation time in the warm, moist tropics, the bug population can go through three generations in the time it takes the rice plant to mature. The number of bugs can explode and overwhelm a field, going from less than one insect per plant to 500–1,000.

Naturally, the first instinct of farmers at the sight of planthoppers in their fields was to bombard them with pesticides. The Indonesians sprayed from the air and from the ground, but the outbreak continued. Over 350,000 tons of rice was lost, enough to feed 3 million people for a year. Many farmers lost nearly everything. Indonesia was forced to become the largest rice importer in the world.

The insect was considered only a minor rice pest prior to the 1970s. What happened that made the brown planthopper a menace? And how did it resist the assault of tons of insecticides?

Careful study of insect growth on farmers' and experimental rice fields revealed an astounding surprise: it was not that plants treated with insecticides had as many eggs, nymphs, and insects on them as untreated plants—they had more! Indeed, insecticide treatment caused up to an 800-fold increase in insect density. This meant that insecticides weren't preventing hopperburn, they were largely responsible for causing it.

How the hell could that happen?

It turns out that many factors were at work. First, the insects had evolved resistance to commonly applied insecticides, such as

FIGURE 8.2 Brown planthoppers on rice plants.

Photo courtesy of IRRI/Sylvia Villareal.

diazinon. But that would only render the insecticide useless. There had to be more to the outbreak, and there was. A second, much more surprising, discovery was that the pesticide actually increased the rate of egg-laying, by about 2.5-fold. And the third factor, well, I am not going to tell you yet the third reason the populations exploded. I am going to save that revelation until after a couple more examples of ecological cancers, for the general rule of regulation that was broken in the rice fields has also been broken elsewhere. Some fields in West Africa are stalked by a much larger pest.

A BABOONIC PLAGUE

In the village of Larabanga, in the savannah of northwestern Ghana, nightfall puts its residents on edge. The rural community of 3,800 is situated just a few kilometers from Mole National Park, home to a diverse array of mammals including hippos, elephants, buffalo, a slew of antelopes and primates, and a variety of cats, from servals to leopards and lions. Villagers commonly encounter wild animals, but it is not the lions that keep them up at night.

Many families meet their needs by farming maize, yams, cassava, and small livestock on shared lands. But in recent years, some very bold four-legged thieves have been working together to slip into the fields under the cover of darkness to raid the crops—olive baboons. In a matter of minutes, a group of a dozen or more animals can strip stands of plants and severely damage many others before slinking away or being chased off by angry farmers.

The baboons have become so brazen that they also scout and attempt to raid the fields during the day. Constant vigilance is required on the part of farmers, who have resorted to using children to guard the precious crops—children who would otherwise be in school. The combined economic and social impact of the marauding primates has precipitated a serious crisis.

Humans and baboons have long lived in close proximity throughout Africa, so when and why did the baboons become such a problem in Ghana?

Part of the answer is found in the handful of protected reserves and parks that have been set aside in different parts of the country. To keep tabs on the wildlife, the Ghana Wildlife Division started keeping a careful census of forty-one species of mammals in 1968. Every month, rangers at sixty-three posts in six different reserves walked six- to nine-mile-long stretches and counted sightings or other signs of each animal. The several-decades-long census is a remarkably detailed accounting of the changes in mammal populations in different sized reserves ranging from the smallest, Shai Hills Resource Reserve (fifty-eight square kilometers), to the largest, Mole National Park (4,840 square kilometers).

The census reveals that of the forty-one species, all but one declined in sightings across the six reserves over the thirty-six-year period from 1968 to 2004, including many species that became locally extinct, especially in the smallest reserves. The one exception? You guessed it—olive baboons, which increased by 365 percent. Moreover, the animals expanded their range in the parks by 500 percent.

I am going to hold off on the question of why the baboons flourished until after describing one more example of an ecological cancer, one that closed a valuable fishery on the Atlantic Coast of the United States.

FAMINE

Bay scallops have long been part of North American culture. Before European settlement, Native Americans along the East Coast harvested the mollusks for their inch-long, white adductor muscles. From the mid-1870s to the mid-1980s, large commercial scallop fisheries operated in Massachusetts, New York, and North Carolina. In 1928, North Carolina led the nation with a harvest of 1.4 million pounds of scallop meat. For many of the state's fishermen, an early winter harvest was an important source of income between other fishing seasons.

But in 2004, the total scallop harvest was less than 150 pounds. The century-old fishery was declared "depleted" and closed for most of the following years, including 2014. Fishermen, state authorities, and scientists asked: What happened?

It was the fishermen who first noticed the clues in their trawl and pound nets, which were pulling up large numbers of cownose rays.

The three-feet-wide fish migrate down the East Coast during the fall and were getting entangled in and damaging the nets. With their poisonous barbs, and no commercial market for the rays, they were becoming a nuisance to the fishermen.

The fishermen complained to University of North Carolina marine biologist Charles "Pete" Peterson, who had been studying the impact of cownose ray predation on bay scallops along the Carolina coast. Peterson teamed up with colleagues at the University of North Carolina and Dalhousie University to study the problem. They found that cownose ray populations had expanded by at least tenfold along the Atlantic coast over the previous sixteen to thirty-five years, to a population of perhaps 40 million animals. Peterson had previously observed how cownose rays wipe out entire populations of scallops at certain locations along the coast. The explosion in cownose rays seemed to explain the absence of scallops from most locations In North Carolina waters. But what explained the greater numbers of cownose rays?

It is time to unravel the mysteries of these cancers.

MISSING LINKS

Microcystis, planthoppers, baboons, cownose rays. What rule or rules of regulation have been broken that enabled these organisms to explode in numbers?

To answer that, we first have to ponder: What could regulate these populations? Elton stressed that if one wants understand the workings of a community of creatures, one should trace the food chain. Could the populations have increased because there is more food available?

For *Microcystis*, that appears to be a good part of the explanation. The element phosphorus is a limiting nutrient for algal growth. The immediate catalyst to algal blooms is the surge of phosphorus (in the form of inorganic phosphate) that enters Lake Erie from farms and other sources in the spring and summer. In the structure of the lake's food chain, the phosphorus exerts a bottom-up effect on algal populations.

But more food does not appear to explain the other cancers. There are plenty of rice plants for planthoppers to feed on in every field, but the bugs usually don't affect very many. And more food would not explain why they boom in the presence of pesticides. Similarly, food would not explain why only baboons increased while all other mammals decreased in Ghanian parks, nor did cownose rays increase because there were more scallops. So, if not food, what else could regulate the numbers of these animals?

Maybe we should look up, not down the food chain from these animals.

That is just what Peterson and colleagues did for cownose rays. Sharks eat the rays, and when the scientists examined records of shark populations along the eastern seaboard, they saw dramatic declines in five species since 1972 including: an 87 percent decline in sandbar sharks; a 93 percent decline in blacktip sharks; and 97–99 percent declines in hammerhead, bull, and dusky sharks. Sharks prey on other animals as well. If the shark declines were responsible for the boom in cownose populations, one would also expect other shark prey to increase in numbers. Sure enough, the researchers found that in addition to the cownose ray, thirteen other prey species, including various small rays, skates, and small sharks increased dramatically in numbers.

A similar explanation accounts for the baboon plague in Ghana. Lions and leopards prey on baboons, and their numbers have plummeted in Ghana's parks, having disappeared completely from three of the six parks by 1986. As the lions and leopards vanished from those parks, the baboons flourished. [Figure 8.3]

Turning to the planthoppers, why did they explode upon treatment of rice fields with pesticide? It turns out that spiders and a few insects are natural enemies of the planthoppers. Wolf spiders and their young spiderlings, for example, are able to consume significant numbers of brown planthoppers and their nymphs, respectively. The pesticide killed off spiders (and other enemies) that keep the insect population under control. In pesticide-treated fields, the predators are reduced, and the pesticide-resistant prey flourishes.

The discovery from these three very different cancers is one simple, common observation: kill the predators, and the prey run amok. The logic of these ecological cancers is familiar. The predators are negative regulators of population growth. Just like a tumor suppressor, they act as brakes on proliferation. Remove these critical links

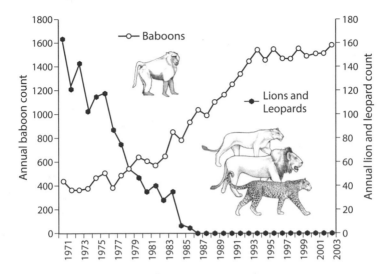

FIGURE 8.3 The increase in olive baboons accompanying the disappearance of lions and leopards in areas of Ghana.

Illustration based on Brashares et al. (2010), redrawn by Leanne Olds.

in the food chain, and the growth of their prey is unchecked, with downstream trophic effects. Each cancer results from the cascading effects of the decapitation of the top trophic level of predators, the reduction of three-level cascades to two. [Figure 8.4]

From the viewpoint of the scallop fisherman, the rice farmer, or a Ghanian family (and in the light of the double-negative logic of the intact cascades), sharks, spiders, and lions should be viewed as allies and not persecuted. In each case, the ancient proverb "the enemy of my enemy is my friend" rings true.

The elimination of trophic levels has probably also contributed to the situation in Lake Erie. In healthy freshwater lakes, algal growth is also controlled from the top down by plankton, such as small crustaceans, which graze on them. In algal blooms, this regulation is overwhelmed or killed off by algal toxins. The algal cancer then is a combination of too much bottom-up input (a stuck accelerator) and too little top-down control (weak brakes).

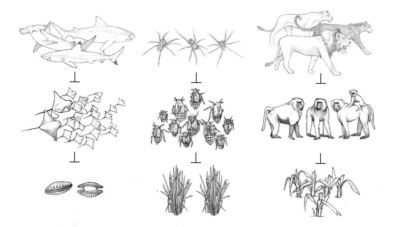

FIGURE 8.4 Cascading effects of the loss of sharks, spiders, and large cats. The loss of control of cownose rays, planthoppers, and baboons has led to a decline in scallops, and loss of rice and other key crops, respectively.

Illustration by Leanne Olds.

TOO MANY, TOO FEW, AND TOO MUCH

Graeme Caughley, who co-wrote a leading textbook on wildlife ecology and management with Tony Sinclair, sorted all the problems of wildlife populations into just three simple categories: too many, too few, and too much. The examples here of too many planthoppers, baboons, and rays are the result of too few spiders, lions, and sharks, respectively.

But the ultimate causes of these cancers are not the missing predators, they are a matter of humans doing too much: too much phosphorus on our farms, too much pesticide on our fields, too much poaching of lions and leopards and their prey, and too much fishing for sharks. It is becoming ever clearer from the kinds of indirect, inadvertent, unanticipated side effects I have described that we are doing these things against our own long-term interest. For many decades, one could say we did not know any better, that we were ignorant of the rules of regulation in nature. But not any longer.

Now that we do know better, can we use our understanding of these rules to fix any of these problem?. Some bold attempts have been and are being made on surprisingly large scales. To quote Cannon when he addressed Boston doctors on the rules of physiological regulation, there are "reasons for optimism in the care of the sick."

FIGURE 9.1 Lake Mendota, Madison, Wisconsin. The University of Wisconsin campus is in the foreground.

Photo by Jeff Miller. Courtesy of University of Wisconsin, Madison.

CHAPTER 9

TAKE 60 MILLION WALLEYE AND CALL US IN 10 YEARS

Good science and good management are hard to do.

—JAMES KITCHELL

I first flew into Madison, Wisconsin, for a job interview one night early in the spring of 1987. From rumors I heard, I did not expect to be offered the faculty position at the University of Wisconsin; and even if I did get it, I did not expect to accept it, as I was anticipating taking a position at a different university. I had never been in the state before and knew little about the place. Since I was unlikely to visit again anytime soon, I decided I would just try to enjoy looking around.

The next morning, I got my first glimpse of Lake Mendota. I was surprised and impressed; the lake was several miles long and across. The university campus hugged about two miles of its southern shore, much of it lined with tall trees, and there were beaches! I had no idea that the university and the city were situated on such a large body of water. Nor did I realize I was looking at what was a famous lake, for no less than Henry Wadsworth Longfellow (1807–1882) had written an homage in 1870 to Madison's lakes (of which Mendota is the largest):

Fair Lakes, serene and full of light,
Fair town, arrayed in robes of white,
How visionary ye appear!
All like a floating landscape seems
In cloud-land or the land of dreams,
Bathed in a golden atmosphere!

I also did not know that I was gazing at one of the best-studied lakes in the world. The lake and the university were the birthplace of limnology (the study of lakes) in North America, pioneering research that began in 1875 when the university was just twenty-five years old and had only 500 (not 43,000) students. Admiring Mendota's calm blue waters and its pretty coastline, I also did not suspect that I was looking at a troubled lake. But like my hometown Toledo's Lake Erie, Mendota was suffering from annual algal blooms and dwindling fish populations.

I sure knew very little that day, including that Madison would become my hometown for the next twenty-eight years. And that very year, Lake Mendota would become the subject of the largest ecological experiment ever undertaken—an application of some Serengeti Rules aimed at curing its ills.

MANIPULATING CASCADES

One of the first truths I learned about my fellow Wisconsinites is that they have a special love for two things: the Green Bay Packers and fish. Almost 90 million fish are taken each year from the waters of the state whose human population is less than 6 million. In fact, the many hardy natives enjoy fishing so much that they actually look forward to venturing out onto the frozen lakes in the dead of winter's subzero temperatures, where many erect custom-furnished shacks, then drill through the thick ice to angle for their favorite quarry—walleye, northern pike, and smaller panfish (bluegill, crappies, and yellow perch).

But in the early 1980s, walleye populations had declined in Lake Mendota to such low numbers that there were few of the fish to pursue at any time of year. Moreover, algal blooms and weeds fouled and choked the shorelines in the summertime. Fishermen and the general public alike urged the state authorities to take action.

The Wisconsin Department of Natural Resources (DNR) tried various ways to improve water quality and fish stocks. Local governments around the lake sponsored programs to reduce the agricultural runoff of phosphorus that fueled the algae and made attempts to remove the weeds mechanically. The DNR supported local fishing clubs in the raising and stocking of walleye. But Mendota is a big lake, and the number of fish that could be raised by the clubs was a drop in a big bucket, and those stocking efforts failed to establish self-sustaining populations. Then James Addis, the director of fisheries of the DNR, read an academic paper that gave him a bold idea.

A few scientists had conducted some novel experiments in a trio of small lakes in the northeastern part of the state, near the Wisconsin-Michigan border. The studies were led by Stephen Carpenter of the University of Notre Dame and James Kitchell of the University of Wisconsin. Early adherents of the trophic cascades concept, Carpenter and Kitchell had proposed that the productivity of lakes was governed by cascades that included up to four trophic levels, including at the top: predators, such as bass, pike, or salmon that ate smaller fish; then the smaller plankton-eating fish; then plant-eating plankton; and at the bottom phytoplankton, such as algae.

To test their hypothesis, they focused on three lakes called Peter, Paul, and Tuesday about two to five acres in size whose food webs were well known. In Peter and Paul Lakes, largemouth bass (predator 1) ate smaller minnows (predator 2), that in turn fed on small crustaceans (zooplankton), that grazed on algae (phytoplankton). An interesting contrast existed in nearby Tuesday Lake, which lacked bass, but where minnows were abundant, plankton were low, and algae grew thick.

From these observations, the scientists predicted that if bass were added to Tuesday Lake, the cascade in the lake would switch to be like that in Peter and Paul Lakes and would cause the abundance of each population at lower levels to invert. To test their prediction, they had to move a lot of fish, almost 400 bass from Peter Lake. Individual bass are hard enough to catch, let alone 400 of them, so they used an electroshocker to bring the fish to the surface and then scooped them up. They were able to remove about 90 percent of the adult bass from Peter Lake and transplant them to Tuesday Lake, and they removed about 90 percent of the minnow population from Tuesday Lake (about 45,000 minnows) to Peter Lake. They left Paul Lake untouched as a control for weather and other factors.

The addition of bass to and reduction of minnows from Tuesday Lake quickly produced the predicted effects. The remaining minnows essentially vanished due to predation by the introduced bass, and the total zooplankton biomass increased by 70 percent. In turn, the population of algae declined by about 70 percent. The addition of the bass and removal of the minnows flipped the cascade in Tuesday Lake. [Figure 9.2]

When Addis learned of these results, it occurred to him that similar biomanipulations might be applied to Lake Mendota to increase its predatory fish populations, reduce its algae, and improve water clarity. Addis phoned Kitchell to discuss the possibility of conducting the same type of experiment on Lake Mendota that he and Carpenter had performed up north. Kitchell was all for it. "Let's do it!" he said.

But Lake Mendota was 2,000 times the area of Peter Lake and much deeper; there were major logistics to figure out. And the lake was also situated just a few blocks from the State Capitol in a populated city; politics had to be considered. And of course, there was also the matter of money.

FRY, FINGERLINGS, AND FISHERMEN

Addis at least had the money figured out. Congress had recently passed an amendment to the Federal Aid in Sport Fish Restoration Act that created a new Aquatic Resources Trust Fund. Paid for by taxes on boat sales, boat fuel, and fishing equipment, support for sport fish programs tripled from $38 million in 1985 to $122 million in 1986. Under the new law, Wisconsin was in line to receive several million dollars more per year. Since one major problem in Lake Mendota was the decline of predatory sport fish, such as walleye and northern pike, a case could be made for using the funding to restore the lake's populations.

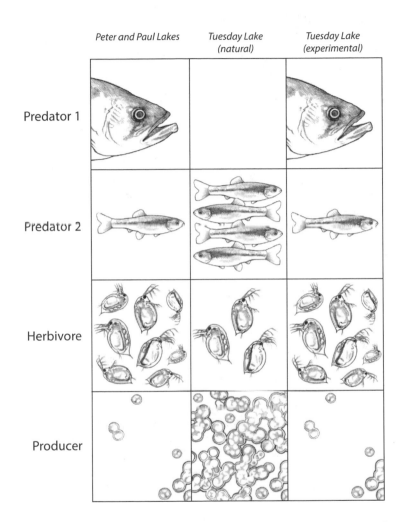

FIGURE 9.2 Manipulation of trophic cascades in Wisconsin lakes. The addition of bass to Tuesday Lake (right) created a similar cascade as those in Peter and Paul Lakes (left), which reduced minnow populations, increased grazing plankton, and reduced algae.

Illustration by Leanne Olds.

But it would not be an easy case. There was the politics of who got the state's fish. Walleye are iconic fish to anglers, worshipped, Kitchell knew, for the simple reason that they tasted great. The state's northern lakes were major tourist draws and needed regular stocking. The DNR's entire walleye production capacity was 3.6 million fish and was insufficient to meet the already existing demand across the state. To ration the fish, the DNR had a policy of limiting the stocking of walleye to 100,000 fingerlings in any one lake, and that number would not make a dramatic impact on walleye populations in Mendota. Kitchell and Addis figured that the lake needed about 25 percent of the entire state's walleye production for several years. That did not go over well with local fishery managers around the state.

Other kinds of political risks threatened the project. To experiment on a large lake in an urban setting, one used by tens of thousands of people, required a great deal of public discussion. And since the experiment was to take place in the state capital, using $1.2 million of public funds, the project was certain to receive considerable scrutiny. The experiment could fail and so confirm some opinions that the project was a waste of money or worse—a waste of fish.

There was also some scientific skepticism, because the trophic cascade concept was not widely accepted at the time. For decades, the prevailing view had been that ecosystems were regulated from the bottom up by nutrients. And unlike the small northern lakes Carpenter and Kitchell had manipulated, Mendota was a so-called eutrophic lake that received large amounts of exogenous nutrients from agricultural and urban runoff, which fueled microbial growth. It was doubted by some whether manipulating the food chain above the microbes could override those inputs. It was also unknown what numbers of fish needed to be stocked to reach sustainable levels, and what those levels would be in the lake. Nor could scientists be sure about how increased walleye or pike populations might affect other species.

One faction that was enthusiastic about the project was the local fishermen and fishing clubs. Indeed, they were so supportive, that they offered to use some their own money to raise fish for the project.

But the scientists knew that fishermen could thwart the entire experiment if they started removing too many fish just as the populations
were trying to expand. So the clubs agreed to endorse highly restrictive new fishing regulations on Mendota, among the most restrictive
of any lake in the state.

Thanks to considerable public support, and Addis's powers of
persuasion in the DNR and the halls of state government, the project
got the green light.

To determine whether the experiment was working, it was essential to know the baseline populations before stocking the lake with
more fish. In the entire 10,000-acre lake, it was estimated that were
fewer than 4,000 adult walleye and 1,400 northern pike. By comparison, the plankton-eating cisco were 200 times more abundant than
either top predator. To boost the numbers of predators, the stocking
of fish began in the spring of 1987.

Now, you might think to increase those numbers, you might add
thousands of fish. Think again. Fish live a Darwinian existence. A
female walleye can lay 50,000 eggs in a single night, but in a stable
population, all but two will perish before adulthood from predation, starvation, and other perils. So, the DNR had to provide a lot of
young fish in the form of what are called fry and fingerlings. Fry are
newly hatched fish that are tiny, about the size of a mosquito, and are
not adept swimmers; fingerlings are juvenile fish that are about two
inches (walleye) or ten inches (pike) in length. The DNR put in 20
million walleye fry and about 500,000 walleye fingerlings into Lake
Mendota each year from 1987 to 1989, more than 60 million fish. They
also stocked 10 million northern pike fry and up to 23,000 fingerlings
each year.

The Lake Mendota project researchers then followed the fate of
the fish and other lake inhabitants, as well as water clarity. Almost
no walleye fry survived, and fingerling survival rates were about 3
percent in the first year after stocking and declined further thereafter.
Still, despite such enormous attrition, the walleye population in the
lake doubled within three years. The larger northern pike fingerlings
fared better than the walleye such that the number of pike over 12
inches in length increased tenfold between 1987 and 1989.

There were some big surprises. The summer of 1987 was unusually hot, and cisco are particularly vulnerable to the increased water temperature and accompanying changes in water chemistry. The hot weather triggered a massive die-off that reduced the initially large cisco population by about 95 percent. This natural event appears to have triggered a trophic cascade in the lake. The cisco feed on lake plankton, such as the small crustacean known as *Daphnia galeata*. The *Daphnia* in turn graze on algae. After the cisco die-off, an extremely rare and larger species called *Daphnia pulicaria* replaced the smaller *galeata*. As it turns out, the larger *Daphnia* graze more intensively on algae and other microbes (phytoplankton). As a result of the cisco die-off, the larger *Daphnia* flourished, the phytoplankton were suppressed, and the lake water was clearer.

Mother Nature appears to have given the Lake Mendota project a little help in that first year, but she did not ruin the experiment. After the first three years, the stocking of fish was reduced to low rates every other year over the next ten years. Over those ten years, the abundance of walleye and northern pike remained stable at about four to six times the level before the experiment. This was despite the fact that fishing activity for walleye and pike increased more than sixfold once the experiment started, thanks to all the publicity about the effort to improve the stocks. Moreover, the cisco remained at low levels; the major grazer was the larger, previously rare species of *Daphnia*; and water clarity was consistently better than prior to the experiment, even though the levels of exogenous phosphorus coming into the lake remained high.

The degree to which the addition of predators changed the lake community is difficult to disentangle from the cisco die-off. It may well be that the increased abundance of walleye and pike suppressed the potential rebound of the cisco and maintained the lake's new state (more predators, fewer planktivores, more plankton, less phytoplankton, and clearer water). Nevertheless, the Lake Mendota experiment was and still is deemed a success. Managers in countries around the world took note, and many lakes have been successfully manipulated by the removal of plankton-eating fish and the addition of top predators.

But it is not just lakes that have been transformed by the stocking of predators.

THE WOLVES AND THE WILLOWS

Around noon on January 12, 1995, US Secretary of the Interior Bruce Babbitt, US Fish and Wildlife Service Director Mollie Beattie, and three others hoisted a large gray steel box from a mule-drawn sled, and carried it over the snow to a holding pen above Crystal Creek, in Yellowstone National Park's Lamar Valley. [Figure 9.3] Babbitt peered into the holes in the box, and saw the gold eyes of a 99-pound female gray wolf blinking back at him. Early that evening, the crate door was opened, and the wolf joined five other members of her pack. Two months later, after the wolves had acclimated to their new surroundings, the door to the holding pen was opened.

For the individual wolves, their release came at the end of a 900-mile odyssey via helicopter, airplane, and horse trailer from their home range in Alberta. For the species, it marked the end of a seventy-year-long journey back to Yellowstone, the last of the predators having been killed there in 1926. And for America's first and most famous national park, it marked the beginning of a new ecological order.

For the humans involved, the wolves' return was also a long, sometimes traumatic journey. The Lake Mendota project was put together, funded, and launched quickly, within two years of its conception, and almost 100 million fish were released with hardly any public opposition or fanfare. The Yellowstone Wolf Restoration project that eventually released just thirty-one wolves into a vast wilderness took more than twenty years, required acts of Congress, had to hurdle lawsuits and court orders, and engendered a massive Environmental Impact Statement that attracted 180,000 public comments.

The scientific rationale was simple. Yellowstone was the American Serengeti. It had the largest concentration of mammals in the lower forty-eight states, and it was the refuge that saved bison and grizzly bears, once numerous across the American West, from extinction. But one species was missing from the more than sixty that flourished

FIGURE 9.3 The first wolf reintroduced into Yellowstone National Park, January 12, 1995. Mollie Beattie, director of the US Fish and Wildlife Service, and Bruce Babbitt, secretary of the US Department of the Interior, carry the cage of the first wolf to be released.

Photo by Jim Peaco. Courtesy of Yellowstone National Park.

in the ecosystem since presettlement times—the wolf. Elk populations irrupted after the wolves were gone, and the larger herds took a heavy toll on the systems trees and plants. The ecosystem was not in its natural state or "intact" without the wolves.

The legal rationale was also simple. In 1973, Congress passed the Endangered Species Act, which required endangered species to be restored, if possible. In 1974, the gray wolf was declared one such endangered species.

The cultural issues, however, were much more complex. The eradication of the wolves accompanied the settlement of the United States, because the predators were deemed a threat to livestock, such as cattle and sheep. The campaign to eliminate them was so intense that the species, once common throughout all of North America, was exterminated in most states by the 1930s. To restore them in just one limited region of the American West was, in many minds, inviting a murderer onto family ranches and farms. Moreover, the large elk herds supported a significant hunting and guiding economy. To add wolves would subtract elk, which would subtract from cash registers and paychecks.

In contrast, wildlife advocates and environmental groups saw the wolf's endangered status as the poster animal for the ignorant and reckless ways in which humans had altered nature, and its reintroduction as a tremendously exciting step forward. But was that sentiment, or science talking? There was skepticism about why it was at all necessary to risk bringing the wolf back where it would certainly feast its way beyond Yellowstone.

The legal prerequisite for settling the competing viewpoints was for the Fish and Wildlife Service to conduct studies that would predict and weigh the benefits and costs of wolf reintroduction—on the wolves, on other wildlife, on hunters, on livestock, as well as on the economy in the greater Yellowstone area. The result was the Environmental Impact Statement (EIS) that presented a variety of predictions based on the projection of an eventual "experimental" population of one hundred wolves living in ten packs.

In terms of wolf prey, the predictions were that elk would be the main prey and that seventy-eight to one hundred wolves would reduce the numbers of northern Yellowstone elk by 5–30 percent, mule deer by 3–19 percent, moose by 7–13 percent, bison by less than 15 percent, and they would not impact bighorn sheep, pronghorn antelope, or mountain goats. For livestock, the EIS predicted little impact of the wolves for the first five years, and over the longer term that the wolves would take an average of nineteen cattle and sixty-eight sheep per year.

Ten years after the release of thirty-one wolves, the population in the greater Yellowstone area was actually 301 animals, but this greater number did not have the expected impact on deer or bison, which remained steady and actually increased, respectively, nor any impact on bighorn, pronghorn, or mountain goats. Confirmed kills of livestock were also within expectations relative to the larger number of wolves, and these kills represent an insignificant fraction of annual sheep (1 percent) and cattle (0.01 percent) losses. The impact on elk, however, was greater: the winter elk population was cut in half between 1995 and 2004, from 16,791 animals to 8,335. On an annual basis, that worked out to ten to twenty elk per wolf.

One major question ecologists pondered prior to the restoration project was what other effects wolves would have beyond their prey. It had been noted long before by pioneering naturalist Aldo Leopold that the disappearance of wolves preceded scores of examples of irruption of deer and elk, which led to the overbrowsing of woody species. As the Yellowstone elk population declined under the resumed pressure of wolf predation, ecologists did start to notice some changes.

Aspen is the most widely distributed tree species in North America and has been declining throughout sectors of the American West. In 1997, biologist William Ripple of Oregon State University noticed that the aspen in Yellowstone were declining. There were several possible explanations, including climate change, fires, insect or other infestations, or overbrowsing. To investigate further, Ripple and his graduate student Eric Larsen took core samples from trees of various sizes across the northern range of the park and counted their annual growth rings.

What they found astonished them. Almost all the trees were at least seventy years old. Eighty-five percent of the trees had matured in the period between 1871 and 1920. Just 5 percent of the trees had originated since 1921. Aspen regenerate by sending off shoots rather than through seeds, and something had prevented those shoots from developing. The clue, Ripple and Larsen thought, was in the age distribution of the trees. Why would aspen have been able to regenerate well up to 1920 but not beyond?

They knew that elk eat aspen, which provide high-quality food and up to 60 percent of the animals' diet in the winter. They also knew that wolves eat elk. But the wolves had been killed off in the 1920s. Ripple and Larsen connected three dots: wolves eat elk, elk eat aspen, therefore the wolves affect aspen. It was a trophic cascade exactly like the sea otter–urchin–kelp cascade in the Pacific. The eradication of the wolves, the keystone species, had unleashed the elk, which in turn suppressed aspen regrowth.

But with the restoration project then under way, Ripple and Larsen suggested, "the re-establishment of wolves on the northern range may be of long-term benefit to aspen." Ripple and other scientists also wondered: if the wolves might benefit the aspen, what other species might be affected?

FIGURE 9.4 Willow recovery after wolf reintroduction in Yellowstone National Park. Photos before (left) and after (right) the reintroduction of wolves show that reduced browsing by elk allows willow (foreground) to flourish.

Left photo courtesy Yellowstone National Park, right photo courtesy William J. Ripple.

Elk also browse on cottonwood and willow that grow along stream banks. Ripple's colleague Robert Beschta began looking at both species in 1996. He noticed that Yellowstone's stream banks were often barren. The willows he did see appeared stunted by heavy browsing. He examined the cottonwood age structure and discovered the same pattern in areas accessible to elk that Ripple had for aspen.

But over the next decade, Ripple and Beschta began to see changes in the aspen, cottonwood, and willow in certain areas of the park. Browsing appeared to be reduced, and young aspen and willows were growing taller, especially near streams. The willows, it turns out, play an important role in the lives of beavers, and vice versa. The shrubs provide food and dam-building materials to the animals; the beavers' dams help provide habitat for the willow. After the wolves' return, the number of beaver colonies in the Lamar Valley rebounded from one to twelve by 2009. [Figure 9.4]

The reintroduction of wolves has also had other cascading effects. Wolves are natural enemies of coyotes, which are smaller *mesopredators*. Coyote populations declined by 39 percent in Yellowstone National Park and in the adjacent Grand Teton National Park at sites colonized by emigrating Yellowstone wolves. Coyotes in turn prey on young pronghorn antelope. Long-term studies have revealed that fawn survival rates were four times higher in sites with wolves than without them.

NECESSITY AND SUFFICIENCY

The reintroduction of wolves has had and continues to exert cascading effects on Yellowstone area communities. But we have to resist being seduced by the idea that the restoration of predators is a panacea for sick ecosystems. Molecular biologists use a couple of handy terms, *necessity* and *sufficiency*, to describe components that are required for a system to work, and those that are able to generate some outcome by themselves, respectively. We have seen ample evidence that predators are often necessary to control prey numbers, but are they sufficient to restore food webs and ecosystem functions?

In the case of Lake Mendota and Yellowstone, the answer is largely "no." In the Lake Mendota experiment, the fortuitous die-off of the plankton-eating cisco was probably a necessary additional condition for triggering the change in plankton and water quality. This is also suggested by manipulation experiments in other troubled lakes, where predator addition was found to also require removal of plankton-eating fish to change the state of the lake. Similarly, researchers in Yellowstone have found that the return of beavers has not occurred in certain parts of the park, where seventy years of erosion and other factors altered the landscape and stream characteristics. Wolves are not sufficient to reverse such physical changes. Putting keystones back helps but may not rebuild the entire arch.

We might expect then that the more altered an ecosystem is, the harder it will be to restore it. But as we shall see next, that does not stop extraordinary people from trying.

CHAPTER 10

RESURRECTION

Politics we shall always have with us, but if wildlife is
destroyed, it is gone forever, and if it is seriously reduced,
its restoration will be a lengthy and expensive business.

—JULIAN HUXLEY

"All over the rest of Africa the battle is on to save game reserves from
the depredations of greed, politics, bad planning, and ignorance.
Here in Mozambique it is already won," proclaimed the November
11, 1970, edition of Durban, South Africa's *The Daily News*. The article
went on to explain the country's ambitious aim to expand the "little-
known gem" of Gorongosa National Park in central Mozambique
into one of Africa's largest game reserves "where Nature can maintain its balance without outside help."

Behind the grand scheme was young South African ecologist Ken
Tinley. The Mozambican authorities hired the doctoral student to
study the park's resources and plan the undertaking. Tinley advocated defining the park's boundaries in such a way that would reflect
and maintain the integrity of its ecosystem. Situated at the southern end of the Great East African Rift, 1,000 miles due south of the
Serengeti, the park got its name from the nearby 6,000-foot Mount
Gorongosa. [Figure 10.2] The mountain's rainforest receives about
eighty inches of rain per year that feeds the rivers winding through

FIGURE 10.1 Gorongosa National Park scene, circa 1960s.

Photo courtesy of Jorge Ribeiro Lume.

the park. The expanded Gorongosa was to include the mountain as well as the different kinds of habitats in which the animals roamed.

Gorongosa had already become a destination for a few of the world's jet set, who came to marvel at its large lion, elephant, and buffalo populations. Tinley conducted the first aerial survey of the roughly 4,000-square-kilometer park in 1972. He estimated populations of 14,000 buffalo, 5,500 wildebeest, 3,500 waterbuck, 3,000 zebras, 3,000 hippos, and 2,200 elephants. The Gorongosa area lion population numbered about 500. Such figures justified the newspaper's faith that "foresight and planning will give Mozambique a reserve probably unmatched in Africa."

After six years of studying the entire Gorongosa ecosystem and making his recommendations for the new contours of the park, Tinley completed his thesis and received his PhD in 1977.

Seventeen years later, Tinley returned as part of a group of scientists who conducted another survey over Gorongosa. Over the course of forty days, they saw no buffalo, wildebeest, or hippos and

FIGURE 10.2 Map of Southeastern Africa.

Drawn by Leanne Olds.

estimated that no more than 129 waterbuck, sixty-five zebras, and 108 elephants remained. The park, whose symbol was a maned lion, had no lions at all. One of the scientists titled his report "A Dream Becomes a Nightmare."

What the hell happened?

PARADISE LOST

What happened was the hell of the Mozambique civil war. After the Marxist Front for the Liberation of Mozambique drove the Portuguese from Mozambique in 1975, a single-party socialist government was established. In an effort to gain control over every facet of Mozambican society, traditional structures were dismantled. Villagers were forced to relocate into towns or communes. Dissidents were placed in "re-education camps" or convicted in show trials, and many were executed. Within two years, these oppressive measures inspired the formation of the Mozambique Resistance Movement (RENAMO).

The conflict that erupted turned into one of the longest, most brutal, and destructive wars in recent decades. Over the course of fifteen years (1977–1992), more than 1 million people were killed in the fighting, thousands were tortured, and 5 million were driven from their homes. RENAMO established its headquarters near Gorongosa, which unfortunately offered the advantages of being situated near the geographic center of the country and of providing refuge and food for the rebels.

As a symbol of the national government, the park was not spared—it was a target. In December 1981, the park's headquarters was attacked; by 1983, the park was closed to visitors and abandoned. RENAMO deliberately destroyed the park's school, post office, and health clinic. Gorongosa was the scene of heavy fighting from 1983 to 1992. Both sides shot wild game for food. Even after a peace accord was signed in 1992, the park continued to suffer from rampant poaching, as there were no rangers to stop it.

In 1995, the European Union funded a project to begin to restore some of the park buildings. The small, seasoned team sent to assess the park was shocked when they walked into the former headquarters

FIGURE 10.3 Gorongosa main tourist camp, 1995.

Photo courtesy of Ian Convery.

and found it in ruins, with bullet holes and graffiti all over the damaged buildings and abandoned vehicles. [Figure 10.3] The park itself was a dangerous place, not because of large animals—they were nearly exterminated—but because of land mines.

With no animals and no facilities, there would be no tourists. The future of Gorongosa was grim. And that is how it remained for several years until, more than 7,000 miles away in Cambridge, Massachusetts, an American businessman learned about the park and dared to wonder: Might it be possible to change Gorongosa's fate, perhaps even to restore it to its former glory?

LOOKING FOR A PROJECT

In 2002, forty-three-year-old entrepreneur Greg Carr was pondering his future. The native Idahoan and Harvard graduate had sold a successful telecommunications business and turned to philanthropy. He started the Carr Foundation in 1999 to focus on human rights, the arts, and the environment. He had endowed a Center for Human Rights Policy at Harvard and launched a new theater company in Cambridge. But Carr was looking for a more "hands-on" project in which to channel his energy, something that might allow him to use his business skills and his wealth to help people.

He began to think a lot about major challenges in Africa, such as the exploding AIDS crisis. Then he had a chat with a Cambridge neighbor, who happened to know Carlos dos Santos, Mozambique's ambassador to the United Nations. "You should go talk to him," his friend said.

Since he had never been to Mozambique, Carr read everything he could about the country before going to New York to meet the ambassador. He learned that it was among the poorest countries in the world. Carr told the ambassador that he was looking for a project that promoted economic and human development. "Come to Mozambique," dos Santos told him, "you can do anything you want." The ambassador was close to President Joaquim Chissano and offered to introduce Carr.

Carr visited the country for the first time in 2002. He was impressed with the beauty of the country, as well as its great size. Twice the area of California, Mozambique has a 1,000-mile-long coastline. Carr also met with President Chissano, who invited him to "adopt a project." But the critical turning point for Carr came not in Mozambique, but when he visited neighboring Zambia and went on his first safari.

Seeing both the beauty of wild Africa and the dire poverty of the people, Carr went back home a changed person. He had thought that his money could be used to build much-needed schools, or health clinics, or drinking wells, but it eventually dawned on him that even if young Africans were able to complete their schooling, there were no jobs for them. He reasoned that to make a difference in Mozambique, he needed to create jobs as well as basic infrastructure. And the obvious industry Mozambique could develop was tourism. Every other East and southern African country had a safari industry, but not Mozambique. Why not? Carr learned about the civil war, and why there was no longer any tourism. But he also heard how tourism had once been an economic engine in the 1960s, particularly in a place called Gorongosa.

Carr decided that rebuilding the tourism industry was his best strategy. He also realized that doing so would require healthy national parks, but up to this point in his life, Carr did not know anything about ecology or conservation. So he began devouring the canon of conservation—Henry Thoreau, John Muir, Aldo Leopold, Rachel Carson, and E. O. Wilson.

With a newfound passion for conservation, Carr returned to Mozambique in 2004 with a list of six potential parks to scout. He brought with him Markus Hofmeyr, the top wildlife veterinarian for the famed Kruger National Park in South Africa, as an advisor. They hired a helicopter from the capital Maputo and began their tour at Limpopo, on the border with South Africa and contiguous with Kruger. They then made their way to Gorongosa.

Carr was struck right away with the physical beauty of the place, its massive rainforest-covered mountain, the Great Rift Lake Urema, and its adjoining wetlands. When asked to sign the park's guest book, he wrote, "This is a spectacular park and it could become one of the best in Africa with some assistance."

Carr had found his project.

In October 2004, he pledged $500,000 toward the park's restoration to Mozambique's Ministry of Tourism. That was to be a small down payment. In November 2005, he agreed to provide $40 million to the park's restoration over a thirty-year period. But this would not be a matter of merely sending checks from the United States. Carr and his foundation were to co-manage the endeavor on the ground with the Mozambicans.

The task before him was herculean. When Carr went back to Gorongosa with a team of engineers, tourism developers, economic advisers, and scientists to begin the work, the main camp was still in ruins, there was barely any running water, and they had only a small generator for electricity. Carr spent the nights in the open bed of a pickup truck. The glory of Gorongosa was a forgotten memory. Mozambicans even told Carr, "Don't bother, there's nothing there anymore."

To find out what if anything was still there, Carr commissioned an aerial survey that was taken in the last week of October 2004. The results were mixed. The spotters did count many more waterbuck, reedbuck, and sable antelope than had been seen ten years earlier at the end of the war. But they saw no zebras, wildebeest, elephants, or buffalo in the survey area—just one lone buffalo outside the grid, and one single lion within it.

When Carr returned home to Cambridge, he was felled by a severe case of malaria—contracted while in his makeshift accommodations. Unable to move, with massive headaches and a recurring high fever, his recovery, let alone Gorongosa's, was uncertain until he found some Boston doctors who knew how to treat the disease that kills one-half million Africans each year.

Neither his state of health nor the dismal state of Gorongosa deterred Carr. The wildlife survey, however, underscored the big question of where and how to start in restoring the park. This was not a case of one or two predators that had been selectively removed. Few places in the world were as broken as Gorongosa: entire chains in the food web had been and remained decimated.

REBUILDING FROM THE BOTTOM UP

While Gorongosa had been famous for its lions—and that's what tourists want to see—the top of the food chain was obviously not the place to start rebuilding, since their prey had been wiped out. Rather, it was clear that the large grazers should be the priority. Their absence had led to major changes in the park's vegetation. Without elephants browsing, woodlands had expanded, and without the large grazers, the grasslands grew tall and fed frequent, intense dry-season fires.

Gorongosa needed animals, but where could Carr find them? The first offer came in from Kruger National Park: 200 disease-free African buffalo. Populations in southeastern Africa were commonly infected with tuberculosis and brucellosis, but Kruger had segregated a population and kept them healthy.

With animals so precious and poaching an ever-present threat, one of the first concerns was where to put the buffalo so that they would be safe and able to thrive. When Carr and Hofmeyr had visited Limpopo, they saw a sanctuary that had been erected in the park. Hofmeyr recommended that Carr do the same for Gorongosa

FIGURE 10.4 First buffalo released, Gorongosa National Park, August 2006.

Photo by Domingos Muala. Courtesy of the Gorongosa Restoration Project.

and picked out a 15,000-acre area for a fully fenced, well-patrolled "Sanctuario" that would keep the new arrivals free from any lions or poachers and give them the best head start in rebuilding the Gorongosa population.

The first batch of fifty-four animals arrived in August 2006. "It was an incredible gift," Carr recalled. But still, the pre-war buffalo population was 14,000. Carr worried that he would need to truck in animals every week for ten years to rebuild the grazer populations. Not only would that be expensive, but the 600-mile journey from South Africa to the park was traumatic for the animals. [Figure 10.4]

Carr soon found out that getting animals, especially the right ones, was going to be much more difficult than that first batch of buffalo. Gorongosa had been home to a distinct subspecies of zebra called Crawshay's zebra, one that has thinner black stripes that extend all the way across their underbellies. The subspecies was once widely distributed across southeastern Africa but had become restricted to a few conservation areas. Carr's scientific team did not want to extinguish the subspecies by bringing in the more common zebra, so they waited several years and worked patiently to get a grand total of fourteen Crayshay's zebras from other provinces in Mozambique, which they also introduced into the Sanctuario. One-hundred-eighty wildebeest were similarly introduced into the Sanctuario in 2007, while six elephants and five hippos obtained from other parks were released directly into the park in 2008, as well as thirty-five eland in 2013.

Now more than ten years since the restoration project began, how are the animals faring? I went to Gorongosa to see for myself.

A GIANT SALAD BOWL

Pilot Mike Pingo picked us up in his bright red five-seat helicopter at Beira's airport, near the coast and about eighty miles southeast of the park (the "us" in this story being myself, my wife Jamie, and Dennis Liu and Mark Nielsen, two colleagues from the Howard Hughes Medical Institute's Department of Science Education). With relatively few roads, and most of those unpaved in this marshy part of the country, the chopper provided a much faster final leg to our twenty-six hour journey, as well as a bird's-eye view of the scenery below.

We passed over miles of flat, largely empty terrain; occasional clumps of trees; and a few small villages, consisting of a handful of thatched huts and small plots of maize and other crops. Then, we started to climb a bit and crossed into the park. My first impression was that Gorongosa was much more heavily forested than I expected, with many tall and short palm trees, and beautiful stands of "fever" trees with green-tinged bark, not at all like the acacia-dominated Serengeti. As we came upon Rio Urema, one of several rivers winding through the park, Pingo started banking left and right over its snaking contours. A few large birds took flight below us, and then I saw *them*.

Crocodiles! Yellow- and black-checkered Nile crocodiles were lining almost every muddy bank or spit. I had seen a good number of crocs before in northern Australia, but never this many nor so close together, twenty or more on a single hundred-foot-long mudbank, including some fifteen-footers. Here was at least one species that was doing well in Gorongosa. The slumbering giants quickly rose as we approached and slid gently into the river. A marvelous sight for this reptile lover.

We then landed at the small airstrip outside the main camp called Chitengo, where Greg Carr and many of his colleagues involved in the Gorongosa Restoration Project (GRP) met us, along with a few warthogs and baboons.

"Would you like to go on a game drive, or would you rather rest after your long trip?" Greg asked (I am breaking form here and will use his first name). It was three in the afternoon; the sun would set in a couple of hours. Adrenaline trumped fatigue, and we were soon in an open-air safari truck bouncing down one of the park's dirt roads. We quickly spotted some impala bounding through the thickets and a few shy oribi (a small antelope). Near several small waterholes, we saw many waterbuck—a large antelope with a distinct white circle on their rear ends. We stopped to admire several birds, including an eagle and a kingfisher.

The trees and tall grass eventually ended at an enormous floodplain. We pulled off the road and climbed out of the truck to take in a breathtaking panorama. The foreground was carpeted with lush green grass and forbs, and was cut by a wide stream and its side-channels, which attracted flocks of yellow-billed storks, whistling ducks, and spur-winged geese. The plants were being grazed by scores of waterbuck and impala. In the background, on the horizon, at the western edge of the Great African Rift, stood massive Mount Gorongosa, whose rainfall swells the waters of Lake Urema, a vast, shallow lake on the floor of the Rift Valley to our north.

"Not a human being, not a light, not a road," Greg noted, as the sun painted a pink-orange backdrop to the Rift. If wowing us upon arrival was his aim—mission accomplished.

***

In the ensuing days, I would appreciate much more about the critical connections between the mountain, the rivers and lake, the floodplain, and the wildlife populations, aided in large part by a helicopter flight over Lake Urema and the floodplain. From the air, I could see that the greatest concentrations of mammals anywhere in the

FIGURE 10.5 Greg Carr, philanthropist and bartender. Greg serving the evening's refreshments from a makeshift bar at Gorongosa's "Hippo House," an abandoned concrete shell overlooking a hippo pool.

Photo by author.

park were on the green floodplain, including large pods of hippos and herds of antelope. The simple question then was: How and why is this so? What made that floodplain?

While the size of the lake was impressive, and the floodplain stretched for miles in all directions away from it, this was Gorongosa's dry season (June). I learned that months earlier, most of the plain was under water, and the lake swollen to many times its present size. The water's retreat during the dry season exposes a huge amount of land that, freshly fertilized with silt and well watered, becomes a giant salad bowl in the valley. That salad feeds a lot of wildlife until the rains return in November.

But of all the mammals in Gorongosa, Greg knew that one species was most important to the recovery of the park and its ecosystems—*Homo sapiens*. The human population surrounding the park is about 250,000, and most subsist on less than one dollar per day. To be successful in the long run, Gorongosa would have to prove to be more valuable as an intact preserve than as farmland, timberland, and larder. Greg was just as eager to show me the many ways in which the GRP was helping the surrounding communities as he was for me to see the wildlife.

HUMAN DEVELOPMENT

In creating jobs for and providing services to the surrounding community, the GRP has, in fact, spent as much money outside the park as inside it. The restoration project provides direct employment to hundreds of area residents. One of the most important jobs for the recovery of wildlife populations is the patrolling of the park to suppress poaching, overfishing, clear-cutting of trees, crop planting, and the illegal setting of fires. One-hundred-twenty armed park rangers, most recruited from the local area, walk for days at a time across the vast wilderness to find and destroy thin wire snares, the most common form of trap that can maim or kill many species. They are also responsible for locating and arresting the poachers who set them. It is difficult, dangerous work. The rangers also have to manage animal-human conflicts, even if that means guarding a villager's crops at night from raiding elephants.

Tourism is another source of jobs and community revenue. To begin to rebuild the tourist industry sixty local scouts were hired in 2006 to guide tourists up Mount Gorongosa. In 2007, new tourist bungalows were opened in the main camp area. A beautiful open-air restaurant was added, as well as a swimming pool. While fewer than 1,000 tourists visited in 2005, 8,000 came in 2008. The policy was established in 2009 that 20 percent of tourism revenues would go to surrounding communities, which are then allocated by local committees to various projects, such as schools, health posts, and fire control.

While Mozambique is poor, the Gorongosa district is among the poorest in the country, and most residents lack access to basic education and healthcare. In 2006, the GRP built a health clinic in Vinho, a community immediately adjacent to the park's southern border, on the other side of the Rio Púnguè. In 2009, a mobile health clinic was launched that provides vaccinations, prenatal and family planning services, and disease prevention tools to surrounding communities. To help curb malaria, mosquito nets were distributed to all 250,000 people in the buffer zone around the park.

The GRP also built a primary school in Vinho in 2006, which previously had none. And in 2010, a Community Education Center was built close to the town of Vila Gorongosa that provides programs for thousands of area children to learn about the local flora and fauna and basic environmental principles, and for area farmers to learn about sustainable agricultural methods. The GRP is providing agricultural expertise to help villagers implement more productive and efficient farming methods and to plant new crops. The simple conservation premise is that if land is used more efficiently outside the park, and food security is improved, there will be less pressure on park lands. On the day we visited the Community Education Center, two dozen farmers were learning about developing coffee crops on the slopes of Mount Gorongosa.

That undertaking may in fact be the most critical to protecting the park's most vital resource—the water that flows from the mountain. Lake Urema and most of the park rivers do not dry up during the long and increasingly hot dry season. They continue to provide habitat and drinking water, because water continues to flow year-round from Mount Gorongosa. Its rainforest acts like a sponge, soaking up rainfall during the wet season, and then steadily releasing it during the long dry season. That flow helps maintain the overall water table in the salad bowl. Water from the mountain, then, is the lifeblood of the Gorongosa ecosystem.

But the mountain's hardwood rainforest has been rapidly reduced by clear-cutting and the planting of food crops, such as maize. Without the trees, the mountain retains less water and the soil washes away. The maize crops do not fare well for long, and then more forest gets cut down. The GRP had to develop some plan to halt and perhaps reverse the destruction of the mountain rainforest. To motivate farmers to behave differently, the GRP had to identify an alternative crop that was more valuable than maize and was compatible with the regrowth of the forest.

The solution? Shade-grown coffee.

The GRP has established an impressive nursery high up the slope of Mount Gorongosa, where we saw over 40,000 potted plants ready to go into the ground. To help the coffee get established, the seedlings are planted alongside pigeon pea, a rapidly growing staple crop that will shade the young plants and provide food and income to the farmers in the meantime. In addition to 2,200 coffee plants, ninety hardwood rainforest trees are being planted on each hectare that will eventually shade each orchard. In the long run, the GRP hopes to reforest a large section of the mountain while establishing a thriving coffee business: the imperatives of conservation met with the imperative of economic and human development.

But the GRP's ambitions have met setbacks. RENAMO has continued as a political and paramilitary organization, and its conflict with the government flared up again in 2013–2014, causing the temporary closure of the park and forcing GRP personnel off the mountain. The parties have signed a new peace agreement, but their political disputes continue.

And then there is the matter of poaching. The top predator of mammals in Gorongosa is not the crocodile or lion, it is humans. In a span of just nine days, I happened to see three pickup trucks carting four or five poachers each to the park's detention center. While poaching in Gorongosa is largely a matter of trapping for meat—and not the slaughter of elephants for their ivory, as in much of East Africa—Pedro Muagura, the director of Conservation, told me that regardless of how many arrests they make, poaching is a relentless, daily toll on park wildlife. Given this reality, I was eager to know what was happening to the numbers of animals, which was impossible to gauge merely from our game drives.

A SALAD, NOW WITH MORE MEAT

On a Saturday night, we were invited to join several scientists and much of the park management as they gathered in an open-air conference room near the airstrip to discuss the state of park wildlife. Early in the genesis of the GRP, Greg was keen to invite the world's scientific community to come study Gorongosa's animals and plants.

Biologists from several universities studying kudu, bushbuck, lions, and even termites were present.

The presentation was kicked off by Marc Stalmans, African wildlife expert and Gorongosa's director of Scientific Services, who had led the most recent aerial survey over the park at the end of the last dry season. Over a span of twelve days, Stalmans and a crew of five flew twenty-nine sorties back and forth across the Rift Valley in Mike Pingo's helicopter (with the doors off), counting all mammals that were large enough to spot from the air—nineteen species of herbivores in all.

The grand total of animals spotted: 71, 086!

The number is particularly astonishing considering that in 2000, there were fewer than 1,000 elephants, hippos, wildebeest, waterbuck, zebras, eland, buffalo, sable, and hartebeest *combined*. Today there are almost 40,000 individuals of these species, including 535 elephants and 436 hippos. Stalmans reported that almost every species they surveyed had increased considerably from recent counts, and most dramatically so from the surveys taken before the start of the restoration project. In just seven years, the density of waterbuck and impala has quadrupled and doubled, respectively, and the number of sable antelope and waterbuck now exceed their historical numbers from the 1960s and 1970s. And all of this has happened despite losses from poaching.

For Greg Carr, the transformation is thrilling. Just ten years ago, visitors could drive all day and see only an occasional animal where now there are herds. He is particularly excited (and relieved) by the fact that he did not have to truck in most of those 70,000 additional animals: nature did most of the work.

"You give nature half a chance, and it's resilient." Greg said. That chance was provided by Gorongosa's park rangers, who have protected the rebounding populations.

That resilience is Serengeti Rule 5 in action (see Chapter 7). The rapidly rebounding populations are exhibiting exactly the same phenomenon that buffaloes, elephants, and wildebeest did in the Serengeti. When populations are at low densities, much lower than their habitat can support, their rates of increase can be very high (see Figure 7.6). Gorongosa's researchers calculate that several species have been increasing at more than 20 percent per year. That is on par with the rapid explosion of wildebeest and buffalo in the Serengeti after the extermination of rinderpest.

Indeed, given adequate protection and habitat, many species across the world have rebounded dramatically from severely reduced numbers. For example, the northern elephant seal was reduced to as few as twenty individuals in the late 1800s and numbers more than 200,000 today, Western Australian humpback whales have rebounded from 300 to 26,000 in fifty years, North Pacific sea otters from 1,000 to 100,000 over the past century, and the American alligator has recovered from near extinction to a population of about 5 million in fifty years.

While the reintroduction of some species may sometimes be necessary, Greg has drawn one lesson from Gorongosa's experience that may guide other large-scale restoration efforts: "Focus on law enforcement, not reintroduction."

But the transformation of Gorongosa is far from complete. Still more grazers are needed, especially buffalo, to cut down on the overgrowth of grasses. And herbivores are not the only important mammals. What about the carnivores? There are no hyenas or leopards in the park; lions are the only large carnivore in Gorongosa. While they have rebounded somewhat from near extermination, they are still at historically low numbers, with just sixty-three individuals tallied at last count. Their scarcity means there is little top-down regulation of prey, which might be contributing to the rapid increase of some of the herbivores. Their rarity would also explain why we had seen every one of the nineteen large herbivore species in the course of several game drives, but not any cats.

One big question the scientists would love to know the answer to is: For how long might animal numbers increase in Gorongosa before leveling off? In the Serengeti, that was when the growth curves turned downward (see Figure 7.6). One fundamental limit to population growth is the total amount of animal biomass that can be sustained by the food and water present in a given area, known as the *carrying capacity* of an ecosystem. Stalmans has estimated that

Gorongosa could support about 8,000 kilograms of animals per square kilometer, with greater densities within the floodplain. The 2014 survey revealed an overall density of about 5,500 kilograms of herbivores per square kilometer, so there appears to be plenty of room for further increases. There is no sign yet that any populations are reaching levels where competition or density-dependent regulation are slowing their increase.

The focus on the numbers of large, visible, and tourist-attracting mammals, while understandable and generally good news, does however tend to overshadow other indicators of Gorongosa's health and importance. That indicator would be the overall biological diversity in the park, and this is where Gorongosa really stands out. Despite all the trauma it has experienced, it is because of the variety of habitats within Gorongosa—rainforest, sand forest, riverine forest, miombo woodlands, limestone caves, grasslands, and floodplain— that it may currently provide home to more species than any other park in Africa. Biologists have estimated there are at least 35,000 species of plants and animals in Gorongosa, including, for example, more than thirty kinds of bats and about 400 bird species. Thanks to our extraordinary guide Fraser Gear, we were able to see a wide variety of residents, including several nocturnal creatures, such as genets; bush babies; various types of mongoose; and my favorite, civets, a beautiful striped and spotted, catlike mammal.

During our last evening in Gorongosa, Fraser took us back to that panorama over the floodplain for one last sunset, then back to camp in the dark. I hoped to see another civet, and Fraser came through as usual. I was then content to just gaze up at Jupiter and Venus in the southern sky, but Fraser stopped, examined some tracks in the road, and pulled off into bush. Sweeping his spotlight back and forth, he rolled us slowly through the tall grass. I had seen the same skill the night before when he tracked a call to a very rare Pel's fishing owl perched over a waterhole. He again made his way to the edge of a waterhole, but when he cast the spotlight to the other side, we saw that he was not tracking an owl. Two lions stared right back at us. A third walked straight at our open-air truck before lying down just twenty feet in front of us. A wider sweep revealed a fourth, a fifth, and still more lions lounging around the bank of the pond, nine in all, including several youngsters. The pride was growing.

Heart pounding, I didn't dare even whisper, but two words came to mind: *Viva Gorongosa*.

FIGURE 10.6 Lions in Gorongosa.

Photo by author.

RULES TO LIVE BY

To save what can be saved, just to make
the future possible: that is the great
motivating force, and the reason for
passion and sacrifice.

—ALBERT CAMUS

We have met the pioneers who discovered the players and rules of some important games, and have seen how that knowledge has been used to advance medicine and conservation. Now, let's take a few moments to gaze ahead at what the future may hold, or more to the point, what it will require. I asserted at the outset of this book that our increasing mastery of biology was a major catalyst to the dramatic changes in the quantity and quality of human life over the past century. What role, then, will biology play in human affairs in the coming century?

On the medical front, I am certain that we still have many miracles to look forward to. We have figured out the drill for combating any disease: identify the most important players, figure out the rules that regulate them, and target what is broken or missing. Our desire for longevity ensures that this cycle of discovery and innovation will continue to be a priority that will be supported by society. For instance, with current generations expected to live into their eighties and beyond, efforts are intensifying to understand and thwart diseases of aging, such as Alzheimer's disease.

But as desirable as such advances may be, they are not the most pressing challenge for biologists and society. Rather, it is the declining health of our home and what that means for the ability of the Earth's ecosystems to support human life, let alone other creatures. We have created the extraordinary ecological situation where we are the top predator and the top consumer in all habitats. As Robert Paine warns, "Humans are certainly the overdominant keystones and will be the ultimate losers if the rules are not understood and global ecosystems continue to deteriorate." Now the only species that can regulate us is us.

If the motto for the twentieth century was "Better Living through Medicine," the motto for the twenty-first must be "Better Living through Ecology." Fortunately, it is not just scientists who are deeply worried or who recognize the primacy of ecology. You may be surprised by who said the following (I certainly was):

> Frequently, when certain species are exploited commercially, little attention is paid to studying their reproductive patterns in order to prevent their depletion and the subsequent imbalance of the ecosystem. . . . Caring for ecosytems requires far-sightedness. . . . Where certain species are destroyed or seriously harmed, the values involved are incalculable. We can be silent witnesses to terrible injustices if we think that we can obtain significant benefits by making the rest of humanity, present and future, pay the extremely high costs of environmental deterioration.

The author is Pope Francis, who issued an extraordinary Encyclical Letter on the environment in June 2015 titled "On Care for Our Common Home." Encyclicals are typically used to address the Catholic hierarchy on major issues of doctrine. In this exceptional instance, the pontiff expressed his wish "to address every person living on this planet" and "to enter into a dialogue with all people about our common home."

The 183-page document is equally remarkable for its ecological scope as for its candor. After systematically assessing the numerous types of degradation, including pollution and climate change, contamination of water, and the loss of biodiversity and species extinction, and their humanmade causes, Pope Francis states:

[W]e need only take a frank look at the facts to see that our common home is falling into serious disrepair. . . . [W]e can see signs
that things are now reaching a breaking point, due to the rapid
pace of change and degradation.

The pope certainly concurs that if we are going to make it into
the next century with sufficient water, productive land and seas, and
the food to support perhaps 10 billion people, as well as preserve
the existence of our fellow creatures, the rules of ecology are going
to have to become rules to live by. Indeed, he calls for a global "ecological conversion," using the latter term in both its spiritual and
everyday connotations as a profound change of heart and mindset.

This is certainly a tall order. Impossible, hopeless many would
say. How can we possibly hope that 7 billion people, in more than
190 countries, rich and poor, with so many different political and
religious beliefs, might even begin to act in ways for the long-term
good of everyone?

Fair question.

But such doubt and pessimism has surrounded other enormous
challenges. One of my motivations in writing this book was to show
how complex, long-term, and even seemingly impossible challenges
have been surmounted. We have dramatically reduced heart disease
and cured leukemias; these were once faraway dreams. The histories
of sea otters, wolves, and other carnivores, and experience from Lake
Mendota, Yellowstone, and Gorongosa show that it is possible to
recover some species, repair habitats, and even resurrect decimated
ecosystems. There is time to change the road we're on.

Our hopes for ecological recovery do have a good scientific foundation, but science, it pains me to admit, is not enough. To have any
power, hope must be strengthened by wisdom about what motivates
action and enables success. And wisdom, as Leonardo da Vinci said,
is the daughter of experience. So we might ask: Is there any precedent
for the sort of global endeavor that biologists and the pope hope for?

I think there is. I will briefly tell the story of another seemingly
impossible but ultimately successful international campaign, one of
humankind's greatest collective achievements, and show how that
experience offers both hope and wisdom for the road ahead.

A SOCIAL UTOPIA

Just fifty years ago, the smallpox virus was still infecting as many as 10 million and killing 2 million people per year. For several years, as mass vaccination eliminated smallpox from numerous countries, the prospect of entirely eradicating the dreaded disease had been discussed. But many thought that was impossible to do, because despite sustained efforts, such as those to vanquish yellow fever and malaria, no disease had ever been completely eliminated from the planet.

Among the most important skeptics was Marcelino Candau, the director-general of the World Health Organization (WHO). The director's doubts were compounded by the fact that the disease existed in at least fifty nine countries. He believed that all 1.1 billion people in these countries (several of which were not WHO members) would have to be vaccinated to stop the disease. The logistics of getting to, let alone vaccinating, 200 to 300 million people per year in hundreds of thousands of villages seemed impossible.

The director had distinguished company. The eminent microbiologist Dr. René Dubos wrote in 1965, "Social considerations make it probably useless to discuss the theoretical flaws and technical difficulties of eradication programs, because more earthly factors will certainly bring them soon to a gentle and silent death. Eradication programs will eventually become a curiosity item on library shelves, just have all social utopias."

But in 1966, under pressure from the Soviet Union as well as the United States, and after three days of debate, the WHO assembly narrowly approved a special budget for smallpox eradication, albeit a paltry $2.4 million. The dismayed director then decided that an American should be put in charge of the program, so that its certain failure would be blamed on the United States and not tarnish the reputation of the WHO.

The general strategy was to vaccinate as many people as possible in smallpox-affected countries. In practice, finding and reaching the last 10–20 percent of the population—including the most indigent, migrants, and remote tribes—was especially difficult. But within a year, an unforeseen breakthrough changed the equation.

Among the American personnel deployed in the field was Bill Foege, a Lutheran missionary doctor working in Nigeria as a consultant to the US Communicable Disease Center (now the Centers for Disease Control). One day in December 1966, Foege received a call on his two-way radio of a suspected case of smallpox in Ogaja Province. Foege and a colleague made their way to the village by motorbike, confirmed the diagnosis, and vaccinated the patient's family and other villagers.

Foege was worried that a wider outbreak might be unfolding. The problem was that about 100,000 people were in the region, and because it was still early in the eradication program, he only had a few thousand doses of vaccine on hand—nowhere near enough to inoculate all the villages.

Foege had to come up with a different strategy. He thought about the "ecology" of the disease, how the disease was spread from person to person and when people were most likely to infect others. He decided that he should focus on vaccinating only those people in places where infections were ongoing. He radioed all the missionaries within about thirty miles and asked them to send runners to every village to check for cases of smallpox. Only four villages had infections; the rest were clear.

Based on information from missionaries in those villages, Foege also identified three places where the patients might have traveled. He and his colleagues then went to each of the infected and potentially infected villages and vaccinated everyone. In this approach, they created a "ring" of vaccination around the infected people, trying to stop the virus from spreading beyond the ring. To Foege, it was similar to techniques he learned as a young firefighter with the US Forest Service: to stop the fire from spreading, remove the fuel ahead of it. Ring vaccination removed the fuel ahead of the virus.

It worked remarkably well. Even though just 15 percent of the population had been vaccinated, within six weeks, the outbreak had been stymied, and no further cases occurred. Their success encouraged Foege's team to extend the strategy throughout eastern Nigeria, where they were able to contain every outbreak while vaccinating far fewer people than mass vaccination would have required.

This experience in West Africa, however, was just a dress rehearsal for the much greater challenge of stopping smallpox in South Asia, particularly in India, where the disease still raged. Foege went to work on the eradication effort in India in 1973. For many years, the country had tried and failed to control smallpox, but their measures were inadequate. The scope of the disease and the size of the population required an effort on the scale of a military campaign. Most cases occurred in four states with large populations, such as Uttar Pradesh (88 million) and Bihar (56 million). The WHO came up with the strategy to visit every village in the country in a span of ten days and then to work immediately to contain any ongoing outbreaks. The initial effort would require an army of 130,000 health workers

Foege and the WHO worked with Indian authorities to train the brigades of health workers who would go to more than 200,000 villages to search out smallpox and then try to contain it by vaccinating around every house where they found it. Before the massive ground campaign began, Prime Minister Indira Gandhi issued a proclamation urging people to cooperate and support the effort. The search teams quickly discovered that the disease was more widespread than they had expected: more than 10,000 cases were spread among 2,000 villages.

The next year, the number of cases soared even higher, with more than 40,000 infected in Bihar alone through April 1974. In May, as the thermometer approached 120 degrees, the working conditions became brutal. With one hundred new outbreaks and 1,000 new cases occurring per day, some Indian officials began to doubt the eradication campaign, which was consuming enormous human resources. But Foege thought the turning point was near. Despite the increasing number of cases and outbreaks, the containment teams were gaining ground, snuffing out over 800 outbreaks per week.

Foege was right. That May turned out to be the worst month. The outbreaks decreased in June and declined steadily through the rest of year. By January 1975, the number of outbreaks had declined to 198; it dropped to 147 in February, with just sixteen new outbreaks detected in the last week of the month; then it dropped again to just three outbreaks in early March. The last case of smallpox in India occurred in May 1975. The scourge that had tormented the country for millennia had been eliminated in just twenty months.

Bangladesh and then Ethiopia were the next to have smallpox eradicated, the latter occurring despite a civil war and WHO teams being kidnapped nine times. The last case of naturally acquired smallpox occurred in Somalia in 1977. The disease was officially declared eradicated in 1980, and smallpox vaccination was no longer necessary across the planet.

Donald A. Henderson, the American who was assigned the unwanted and supposedly doomed task of leading the campaign in 1966, said, "It was a victory achieved by a great many dedicated, imaginative people who did not know that smallpox eradication had been deemed by many professionals to be an impossible goal."

Smallpox is the only human disease to have been eradicated, but it was not a one-off miracle. In 2011, another pestilence that once caused famines, ruined cultures, and toppled civilizations was banished from the world. After decades of effort, the rinderpest virus, the scourge of not just the Serengeti but also of ancient Egypt and the Roman Empire, the cause of cattle plague in Europe and England in the mid-1800s, and most recently a threat across modern Africa and Asia, was declared eradicated ten years after the last infection occurred in Kenya.

The campaign that defeated rinderpest was similar in many ways to the smallpox story. Early attempts were unsuccessful, because affected populations were difficult to reach. After the disease came roaring back in an epidemic that struck eighteen countries across sub-Saharan Africa, new strategies involved training platoons of community animal-health workers who were familiar with local customs and geography. These workers were then able to reach and vaccinate pockets of cattle populations in areas that still harbored the virus—many in troubled areas, such as Ethiopia, Somalia, Sudan, Pakistan, and Yemen.

These two campaigns completely eliminated viruses that infected millions of people and animals across scores of countries, many in remote villages that lacked any medical or veterinary infrastructure or education, and often in regions with ongoing political and civil

unrest. Beyond the immense logistical challenges, the tactics sometimes entailed unpopular measures, such as the quarantining and
guarding of infected houses or the destruction of infected animals.
What factors enabled the campaigns to succeed? And what might
those tell us about how to approach the changes needed to implement more ecologically sustainable practices?

LESSONS TO LEARN FROM

In 2011, Bill Foege published a memoir of his experiences in the smallpox campaign. In his conclusion to *House on Fire*, he offered eighteen
lessons that he thought were applicable to other public health projects. I was struck by how well these lessons applied to tackling any
difficult goals for the common good. I have selected a handful of
Foege's most important lessons (set in italics) and elaborate on their
connections to ecological aims and examples.

Global efforts are possible. Foege emphasizes that global campaigns
materialize around shared risks, such as nuclear arms during the
Cold War or the HIV pandemic, and the identification of common
objectives. The present ecological risks pose the same magnitude of
threat.

Smallpox eradication did not happen by accident. Foege explained,
"This is a cause-and-effect world, and smallpox disappeared because
of a plan, conceived and implemented on purpose by people. Humanity does not have to live in a world of plagues, disastrous governments, conflict, and uncontrolled health risks. The coordinated
action of a group of dedicated people can plan for and bring about
a better future. The fact of smallpox eradication remains a constant
reminder that we should settle for nothing less."

The ecological situations we face are also matters of public health
and welfare; humanity cannot live in a world of polluted water and
air, algae-choked lakes, empty seas, and denuded lands, so we cannot
resign ourselves to such a world.

Coalitions are powerful. One characteristic of successful coalitions
that Foege highlights is the suppression of individual egos for the
sake of achieving a common goal. Turf battles and competition
among individuals, teams, and organizations (public and private) for

resources, for power, or for credit must be suppressed by an unwavering focus on the common goal.

One coalition that I would like to see emerge is the entire biological community standing together to support ecological priorities. This means the molecular tribe, a very numerous and influential group, must get on board and use that influence to support ecological research and education.

The pope's Encyclical also shows that new kinds of alliances are possible among groups that have historically not been allies. "If we are truly to develop an ecology capable of remedying the damage we have done, no branch of the sciences and no form of wisdom can be left out, and that includes religion."

Social will is crucial, and must be transformed into political will. "Government support for programs depends on the agreement of the governed," Foege noted. We have known for decades how to take better care of the environment and natural resources, and many important laws have been enacted (Convention on International Trade in Endangered Species; US Endangered Species Act; US Marine Mammal Protection Act) that have enabled the recovery of dozens of species.

Every scientific recommendation requires political action for implementation. Scientists must equip politicians with the information necessary for making good public policy. I would add that another approach to securing the necessary political will is for scientists themselves to pursue public office.

Solutions rest on good science, but implementation depends on good management. The success of the smallpox eradication effort, which was not well funded, depended on excellent management of resources and personnel. Much of that is credited to Donald Henderson. At a meeting in Kenya in 1978, the then director-general of the WHO turned to Henderson and asked him what was the next disease to be eradicated. Henderson reached for the microphone and said that the next disease to be eradicated is bad management.

The stories of Lake Mendota, Yellowstone, and Gorongosa are the products of good management by the Wisconsin Department of Natural Resources, by the US Fish and Wildlife and National Park Services, and by the joint Gorongosa Project–Mozambique National

Parks team, respectively. The depletion of fish and wildlife, the deterioration of water quality, and the like, are often matters of poor management, not lack of knowledge.

The objective may be global, but implementation is always local. Local cultures and needs determined which tactics were most useful in the smallpox campaign. Similarly, in terms of preserving and restoring nature, the global effort will be the sum of countless local initiatives.

In Indonesia, for example, once the hazards of the misapplication of pesticides to rice crops was understood, a massive campaign was undertaken to establish Farmers' Field Schools and to educate over 1 million farmers in integrated pest management, so that natural predators were allowed to control brown planthopper populations. The government also banned a large number of pesticides.

Be optimistic. Foege says, "The trouble with being an optimist is that people think you don't know what is going on. But it is the way to live." He cautions that there is a place for pessimism, "but don't put those people on your payroll."

When Greg Carr was asked for his motto, he replied, "Choose optimism because the alternative is a self-fulfilling prophecy."

The measure of civilization is how people treat one another. Foege states that the smallpox vaccination program was a "civilized program" not only because it protected vaccinated people but also those around them, as well as unseen people in the generations to come.

The pope expressed a similar sensibility when he asked, "What is the goal of our work and all of our efforts? . . . We need to see that what is at stake is our own dignity. Leaving an inhabitable planet to future generation is, first and foremost, up to us. The issue is one that dramatically affects us, for it has to do with the ultimate meaning of our earthly sojourn."

To this worldly wisdom, I will add three points about the ecological challenges before us. The first is that nothing gets conquered everywhere at once. Both smallpox and rinderpest were eradicated in some countries long before others. Gorongosa is just one park in one country on a continent in which lions have disappeared from twenty-six countries. Progress is important wherever it can be made, and we can't be frustrated into paralysis by a lack of global action.

This leads to my second point: important challenges can't wait until everyone gets on board. The WHO director opposed the small-pox eradication program and even UNICEF refused to fund it.

And finally, individuals' choices matter. Most of us do not have Greg Carr's or the WHO director's resources, but the important point is what they chose to do, or not to do, with those resources. But regardless of resources, everyone has some choice about how to contribute. And that brings me to one final story.

On October 12, 1977, two very sick children with smallpox were brought to a hospital in Merca, Somalia, near Mogadishu, and needed to be taken to an isolation area. Ali Maow Maalin, a kindly twenty-three-year-old hospital cook, volunteered to show the driver where to go and climbed in to the back of the Land Rover next to the children for the short ride.

Two weeks later, Maalin developed a fever and a rash that was first diagnosed as chicken pox. Maalin walked around the hospital be-fore finally being recognized as having smallpox. Vaccination against smallpox had been mandatory for all hospital employees, but on the day Maalin was to receive the vaccine, he was too scared. "It looked like the shot hurt," he admitted later.

Maalin's case prompted the WHO to call in epidemiologists working across Somalia to help contain the disease. The hospital was closed to new admissions, all existing patients were put under quar-antine, and a twenty-four-hour guard was posted. Everyone who had any contact with Maalin was tracked down and monitored closely. All individuals from the houses and the entire ward surrounding Maalin's home were vaccinated. Checkpoints were also set up, so that everyone entering or leaving Merca could be stopped, checked, and vaccinated.

The effort paid off. No new cases appeared in Somalia, or any-where else for that matter. Ali Maow Maalin was the very last person in the world to catch smallpox.

Thirty-one years later, after an intensive eradication campaign in-volving more than 10,000 volunteers and health workers, Somalia was also declared free of polio. One of the workers who criss-crossed the country for several years as a WHO community coordinator, encouraging parents to have their children vaccinated, was Ali Maow Maalin. "I tell the community that I am the last person who [had] smallpox," he explained. "Somalia was the last country with smallpox. I wanted to help ensure that we would not be the last place with polio too."

ACKNOWLEDGMENTS

This book was born on the Serengeti and completed in Gorongosa.

For someone born in the biodiversity "coldspot" of Toledo, Ohio, and who dreamed of going to faraway places with wild animals, I am amazed by my good luck. And to be able to share these special places and experiences with the people I love, well, that is just more than anyone could wish.

Special thanks to my wife Jamie, who braved the uncertainties of Africa and is now ready to move there; to our sons Will, Patrick, and Chris, and to Kristen Finkbeiner for being the best safari companions; and to Moses, our extraordinary guide. Asante sana!

Also very special thanks to Greg Carr and our friends at Gorongosa: Mike Pingo, Vasco Galante, Mike Marchington, Marc Stalmans, Fraser Gear, Sandra Schonbachler, Pedro Muagura, Mateus Mutemba, Paola Bouley, Rob Pringle, Ryan Long, Corina Tarnita, Matt Jordan, Tara Massad, and James Byrne for their hospitality, expertise, and dedication. Thanks also to Ambassador Doug Griffiths and Alicia, Claire, and Helen Griffiths for sharing such memorable game drives and sunsets.

This book would not have been possible without the cooperation and efforts of many people. My gratitude to Joe Goldstein, Ed Skolnick, Roy Vagelos, Robert Paine, Tony Sinclair, Jim Kitchell, Steve Carpenter, Jim Addis, and Greg Carr for generously giving their time to be interviewed about their work, and for supplying me with documents and photos.

For those I was not able to interview, I received great assistance from archives, family members, libraries, and librarians in unearthing their stories. Thanks to Nina Balter at the Howard Hughes Medical Institute for reaching out to other libraries and tracking down numerous critical documents. Thanks to Ivar Stokkeland of the Norwegian Polar Institute library and the Bodleian Library at Oxford University for materials on Charles Elton; to the Countway Library of Medicine at Harvard Medical School for materials on Walter Cannon; to the Archives of the Pasteur Institute and Olivier Monod for photographs of and by Jacques Monod; and to David Rowley for recollections of his mother Janet. Thanks also to Nick Jikomes for special library services in Boston.

Huge thanks to Megan Marsh-McGlone, who has assisted me throughout the research and writing process, tracked down countless sources, curated and assembled the bibliography and notes, secured permissions for all illustrations, and prepared the manuscript. Special thanks also to Leanne Olds, who drew many of the illustrations and prepared all figures for production.

I am also grateful to Dennis Liu, David Elisco, John Rubin, Laura Bonetta, Anne Tarrant, and Jamie Carroll for detailed feedback on drafts of the book, and to Andrew Read, Harry Greene, and Simon Levin for their thoughtful, expert reviews.

Special thanks are also owed to Alison Kalett, my editor at Princeton University press. The spark for this book was an unusual invitation and a challenge from Alison to write a "short, provocative" book on biology. I hope she got what she wanted. And thanks to my agent Russ Galen for his wise counsel and support, and for making sure I got what I wanted.

NOTES

FRONTMATTER EPIGRAPH

Page

vii **Suppose it were perfectly certain:** "Professor Huxley's Hidden Chess-Player." *Spectator Archive*, January 11, 1868. http://archive.spectator.co.uk /article/11th-january-1868/9/professor-huxleys-hidden-chess-player.

INTRODUCTION

Page

4 **Scourges such as smallpox:** World Health Organization (2011) "Chapter 1: Smallpox: Eradicating an Ancient Scourge." In *Bugs, Drugs and Smoke: Stories from Public Health*. Geneva: World Health Organization. http:// www.who.int/about/history/publications/public_health_stories/en/.

4 **Similar gains occurred for malaria:** Prüllerg and Quartburg (2000), p. 34.

5 **Every kind of molecule in the body:** Anderson and Anderson (2002).

6 **Indeed, the majority:** King, S. (2014) "FirstWord Lists—Pharma's 50 Biggest Selling Drugs: AbbVie's Humira Joins the $10 Billion Club." *FirstWord Pharma*, March 7, 2014. http://www.firstwordpharma.com/node /1194000#axzz37MHaRznr.

7 **The criteria for approval:** Ledford (2011).

8 **Tanzania now holds 40 percent:** Riggio et al. (2013).

8 **Now, 26 percent:** Dulvy et al. (2014).

9 **As the authors:** "World Footprint." *Global Footprint Network*. http:// www.footprintnetwork.org/en/index.php/GFN/page/world_footprint/.

10 **"a near infinitude of particulars":** Stent (1985), p. 2.

CHAPTER 1

Page

15   **"The living being":** Cannon (1929), p. 399.

17   **The hypothalamus also:** Everly and Lating (2013), Chapter 2; "Understanding the Stress Response." *Harvard Mental Health Letter*, March 2011. http://www.health.harvard.edu/newsletters/Harvard_Mental_Health _Letter/2011/March/understanding-the-stress-response.

17   **Coined and described:** Cannon (1927).

18   **In a classic series of studies:** Benison, Barger, and Wolfe (1987), p. 56.

18   **Noticed sev times:** Benison, Barger, and Wolfe (1987), p. 62.

18   **"It has long been common knowledge":** Cannon (1898), p. 381.

18   **His talent, rigor:** Benison, Barger, and Wolfe (1987), pp. 70–71.

19   **Cannon knew that:** Cannon (1927), pp. 26–27.

19   **These results showed that the inhibition:** Cannon (1909); Cannon (1911a), p. 219.

19   **Cannon wondered:** Cannon (1927), pp. 40–41.

19   **"made use of the natural enmity":** Cannon (1927), p. 44.

21   **[Cannon] discovered that the blood:** Cannon and de la Paz (1911).

21   **This was the same:** Cannon (1927), pp. 49–54; Benison, Barger, and Wolfe (1987), p. 62.

21   **epinephrine sped up heart rate:** Cannon (1927), pp. 54–57.

21   **the release of sugar:** Cannon (1927), pp. 69–72.

21   **and even blood clotting:** Cannon (1927), pp. 306–308.

21   **Cannon suggested that the responses:** Cannon (1914).

21   **"The organism which":** Cannon (1914), p. 372.

21   **Cannon's student Philip Bard:** Finger (1994), pp. 285–286.

22   **Cannon recognized some of the shock symptoms:** van der Kloot (2010).

22   **"Are there not untried":** Benison, Barger, and Wolfe (1987), p. 390.

22   **Cannon said goodbye:** Cannon (1873–1945), p. 3.

22   **While ships usually have lights:** Cannon (1873–1945), p.12.

22   **The appearance of a British destroyer:** Cannon (1873–1945), p.14.

22   **Although Cannon had not:** Wolfe, Barger, and Benison (2000), p. 9.

23   **Cannon proposed a simple:** Wolfe, Barger, and Benison (2000), p. 14.

23   **"Well, on Monday":** Wolfe, Barger, and Benison (2000), p. 15.

23   **Cannon described three other soldiers:** Wolfe, Barger, and Benison (2000), p. 15.

24   **Cannon and the Allied medical:** Wolfe, Barger, and Benison (2000), p. 18.

24   **"the most stupendous":** Cannon (1873–1945), pp. 396–397.

24   **Then came a flood:** Cannon (1873–1945), p. 401.

24   **As the shock ward filled:** Cannon (1873–1945), p. 398.

25   "There is satisfaction now": Cannon (1873–1945), p. 483.
25   In a span of: Wolfe, Barger, and Benison (2000), p. 535.
25   "For exceptional meritorious": Mead (1921), p. 157.
25   After a joyous celebration: Wolfe, Barger, and Benison (2000), p. 63.
25   He knew too well: Cannon (1929).
25   Cannon began to speak: Wolfe, Barger, and Benison (2000), p. 162; Cannon (1928).
27   "the organism automatically": Cannon (1928), p. 593.
27   Cannon first elaborated: Cannon (1929).
27   then in a popular science book: Cannon (1963).
27   *The Wisdom of the Body*: Cannon borrowed the term from Ernest Starling; it was the title of Starling's 1923 Harveian Oration. Starling (1923).
27   He offered several lines: Cannon (1929); Cannon (1963).
27   "regulation in the organism": Cannon (1929), p. 427.
28   Some compared it to Darwin's: Fleming (1984).
28   "That you, a group of physicians": Cannon (1928), p. 593. Subsequent quotes are from the same source.

CHAPTER 2

Page
30   "[T]he study of the regulation": Elton (1924), p. 154.
30   Elephants are prodigious: "Feeding Ecology and Diet." *Nature Online.* NHM website, Natural History Museum, London. http://www.nhm.ac .uk/nature-online/species-of-the-day/biodiversity/endangered-species /loxodonta-africana/feeding-diet/index.html.
30   Females do not mature: "Feeding Ecology and Diet." *Nature Online.* NHM website, Natural History Museum, London. http://www.nhm.ac .uk/nature-online/species-of-the-day/biodiversity/endangered-species /loxodonta africana/feeding-diet/index.html.
31   The elephant is reckoned: Darwin (1872), p. 31. Note: This statement in the sixth edition differs from the version in previous editions. Alerted by a reader, Darwin realized that his original initial arithmetic was faulty: Darwin (1869, June 26), Letter "Origin of species [On reproductive potential of elephants]." *Anthenaem* 2174: 861. Web. *Darwin Online.* http://darwin -online.org.uk/content/frameset?viewtype=text&itemID=F1746& pageseq=1.
32   "Population, when unchecked": Malthus (1798), pp. 4–5.
33   It was a bold and ambitious: Binney (1926), p. 23.
33   He began a diary: Elton, C. S. (unpublished) *Memoir for Royal Society,* Oxford University Bodleian Library Special Collections and Western Manuscripts, Folder A.36.
33   He had relatively little time: Elton (1983), p. 5.

33   **Elton was, by his own account:** Elton (1983), p. 6.

34   **"I must do something":** Elton (1983), p. 13.

34   **With nothing in his stomach:** Elton (1983), p. 13; Longstaff (1950), pp. 237–260.

34   **"I must have my eight hours":** Elton (1983), p. 15.

34   **"are doing behind the curtain":** Stolzenburg (2009), p. 8.

35   **Among the most numerous:** Summeryhayes and Elton (1923); "Bjørnøya." *Norwegian Polar Institute.* http://www.npolar.no/en/the-arctic /svalbard/bjornoya/.

35   **Summerhayes was as keen:** Elton (1983), p. 5.

35   **He preserved some aquatic animals:** Elton (1983), p. 9.

36   **"stew of fulmar":** Elton (1983), pp. 23, 30.

36   **It was a brutal seven-mile:** Elton (1983), pp. 27–28; Longstaff (1950), p. 241.

36   **"is rather fun":** Elton (1983), p. 30.

36   **After four days:** Gordon (1922), p. 12.

36   **A whale's spout:** Elton (1983), p. 34; Gordon (1922), pp. 19–27.

37   **Elton had learned that Longstaff:** Elton (1983), p. 41.

37   **Shivering with hypothermia:** Elton (1983), p. 53.

37   **To learn more about how:** Elton (1983), pp. 99–100.

38   **But now that it had found:** Elton (1983), p. 92.

38   **After more than two months:** Elton, C. S. (unpublished) *Small Adventures*, Unpublished Manuscript, Oxford University Bodleian Library Special Collections and Western Manuscripts, folder A.32, p. 39.

39   **Elton drew a schematic:** Summerhayes and Elton (1923).

40   **Binney would also organize:** Barker (2005).

40   **The ship was thwarted:** Binney (1926), pp. 24–29.

40   *Norges Pattedyr* **by a Robert Collett:** available at Biodiversity Heritage Library online, http://www.biodiversitylibrary.org/item/51536#page/9 /mode/1up.

40   **Although Elton did not read Norwegian:** Elton, C. S. (unpublished) "How Are the Mice," *Small Adventures*, Unpublished Manuscript, Oxford University Bodleian Library Special Collections and Western Manuscripts, folder A.32, p. 1.

41   **There were fifty or so pages:** Elton, C. S. (unpublished) "How Are the Mice," *Small Adventures*, Unpublished Manuscript, Oxford University Bodleian Library Special Collections and Western Manuscripts, folder A.32, p. 1.

41   **Like Archimedes in his bathtub:** Elton, C. S. (unpublished) "How Are the Mice," *Small Adventures*, Unpublished Manuscript, Oxford University Bodleian Library Special Collections and Western Manuscripts, folder A.32, p. 2.

41   **Elton noted that short-eared owls:** Elton (1924), p. 132.

42   "It lives on Rabbits": Elton and Nicholson (1942), p. 241.
42   Elton documented these: Elton (1924).
43   "chiefly concerned with": Elton (1927), p. vii.
43   "It is clear that animals": Elton (1927), p. 55.
43   "subject to economic laws": Elton (1927), p. viii.
43   The concept of an "economy of nature": Pearce (2010).
43   "At first sight": Elton (1983), p. 55.
44   "Food is the burning question": Elton (1927), p. 56.
44   "The large fish eat the small fish": Elton (1927), p. 50.
44   In Elton's scheme: Elton (1927), p. 56.
44   "Large fowl cannot eat": Elton (1927), p. 50.
44   "the size of the prey": Elton (1927), p. 60.
44   "One hill cannot shelter": Elton (1927), p. 50.
45   There was generally a progressive: Elton (1927), p. 69.
45   "vast numbers of small herbivorous": Elton (1927), p. 69.
45   "all over the world": Elton (1927), pp. 69 70.
45   The pyramid of numbers: Elton (1927), p. 113.
45   Elton suggested that, in general: Elton (1927), pp. 122–123.
45   "It has been necessary": Elton (1927), p. 120.
46   "when a whole lot of lemmings": Southwood and Clarke (1999), p. 137.
46   "The lemmings march chiefly": Elton (1927), p. 133.
46   The scene was faked: "The Lemming—A Misunderstood Rodent." BBC
      Online. http://www.bbc.co.uk/dna/place-lancashire/plain/A43520645.

CHAPTER 3

Page

51   "The cell thus adapts": Jacob (1973), p. 282.
51   Braving ice, storms, temperatures: "Jean-Baptiste Charcot 1867–1936."
      South-Pole.com. http://www.south-pole.com/p0000096.htm.
51   This voyage was the sixty-seven-year-old's: Vibart, E. (2011) "Jacques
      Monod à bord du Pourquoi Pas?" Voiles et Voiliers, December: 234–243;
      Debré (1996), p. 69.
52   Also on board: Among the four was Paul Émile-Victor, who would be-
      come a very well-known author and explorer.
53   For five days: Vibart, E. (2011) "Jacques Monod à bord du Pourquoi-Pas?"
      Voiles et Voiliers, December: 234–243.
53   "I saw so many beautiful": Vibart, E. (2011) "Jacques Monod à bord du
      Pourquoi-Pas?" Voiles et Voiliers, December: 234–243, p. 242.
53   Monod published a preliminary account: Drach and Monod (1935).
54   The crippled ship drifted: Lemarchand, F. (2011) "Comment Disparu-
      rent le Pourquoi-Pas? Et Son Équipage?" Omniscience, October 27: 62–63.
      http://www.omniscience.fr.

54   **André Lwoff, a microbiologist:** Lwoff (2003), p. 4.

55   **One of his first clear-cut results:** Monod (1942), figure 5, p. 38.

55   **On August 31, 1939, Monod:** Letter, Jacques Monod to his father and mother, 8/31/1939, Private archives Monod family, qtd. in Carroll (2013), pp. 37–38.

56   **"That could have something":** Carroll (2013), p. 134.

56   **"Enzyme adaptation?"** Carroll (2013), p. 134.

57   **To test this idea:** Monod (1942), p. 167.

57   **"What Monod is doing":** Lwoff (2003), p. 5, quoted in Carroll (2013), p. 135.

57   **Monod hoped to study:** Carroll (2013), pp. 163–176.

57   **hiding incriminating documents;** Carroll (2013), p. 183.

61   **This turning on and off of enzyme production:** Hogness, Cohn, and Monod (1955).

65   **Jacob and a visiting American scientist Arthur Pardee:** Pardee (2003), p. 112.

65   **Maybe the inducer did not activate:** Judson (1979), pp. 408–410.

66   **Until, in the darkened cinema:** Jacob (1988), pp. 297–298.

66   **A repressor must also:** Jacob (1988), pp. 297–298.

67   **In addition to breaking down nutrients:** Abelson, Bolton, and Aldous (1952), pp. 173–178.

67   **However, when specific amino acids are provided:** Abelson, Bolton, and Aldous (1952), pp. 173–178.

67   **It was discovered, for example, that when the amino acid tryptophan:** Novick and Szilard (1954).

67   **Similarly, it was found:** Umbarger (1956).

68   **These discoveries inspired:** Umbarger (1961).

69   **Late one evening in the fall of 1961:** Ullmann (2003), p. 201.

71   **"society of macromolecules":** Jacob and Monod (1963), p. 1.

72   **"well-known axiom":** Monod and Jacob (1961), p. 393.

72   **"On the other hand":** Monod and Jacob (1961), p. 396.

72   **Recognizing that cancer cells:** Monod and Jacob (1961), p. 400.

CHAPTER 4

Page

73   **They remained above 20,000 feet:** Hoffman, W. (1979), "Meet Monsieur Cholesterol." *University of Minnesota Update, Winter 1979.* http://mbb net.umn.edu/hoff/hoff_ak.html; Echevarria, E. (1986) "Early American Ascents in the Andes, 1854–1950." *American Alpine Journal* 28(60): 111. http://publications.americanalpineclub.org/articles/12198611100/Early -American-Ascents-in-the-Andes-1854–1950.

73   **The expedition was the largest:** Tracy (2012).

74   **A gifted child growing up in California:** Keys, A. (1961) "The Fat of the Land." *Time Magazine* 77: 48–52.

75   **So, the Quartermaster Corps:** Keys (1990).

75   **After further trials at Fort Benning:** Keys (1990).

75   **Dubbed the K-ration (purportedly for "Keys"):** Hoffman, W. (1979), "Meet Monsieur Cholesterol." *University of Minnesota Update, Winter 1979.* http://mbbnet.umn.edu/hoff/hoff_ak.html; although see B. F. Shearer (2007) *Home Front Heroes: A Biographical Dictionary of Americans During Wartime.* Vol. 3. Westport, CT: Greenwood Press, p. 477.

75   **Keys recruited 281 Minneapolis-area men:** Keys et al. (1963).

75   **Studying a group of Neapolitan firemen:** Keys (1990).

76   **Men with levels greater than 260 milligrams:** Keys (1990).

76   **The Seven Countries study:** "Cross-Cultural Correlations." *The Seven Countries Study.* http://sevencountriesstudy.com/study-findings/cross -cultural.

76   **It had been known for fifty years.** Goldstein and Brown (2003).

77   **After long days on the wards:** Williams (2010).

78   **Parts of their very full days:** Joseph Goldstein, Phone Interview, 11/20/2014.

78   **Goldstein and Brown had learned:** Gould et al. (1953).

78   **"It's just an amorphous glob"** Williams (2010), p. 1010.

79   **One of the first things:** Brown, Dana, and Goldstein (1973).

79   **They discovered that LDL:** Brown, Dana, and Goldstein (1973).

79   **When cells from FH patients:** Goldstein and Brown (1973).

79   **They found that cholesterol was indeed a potent inhibitor:** Brown, Dana, and Goldstein (1974).

80   **They observed that labeled:** Goldstein and Brown (1974).

80   **Brown and Goldstein also discovered:** Brown and Goldstein (1975).

81   **More than 93 percent of all:** Brown and Goldstein (1981).

81   **In college, he learned:** Landers, P. (2006) "Stalking Cholesterol: How One Scientist Intrigued by Molds Found First Statin." *Wall Street Journal,* January 9; Rea (2008).

81   **He identified a parasitic fungus:** Endo (2010).

81   **And indeed, many drugs:** Endo (2010).

82   **The second, isolated in the summer of 1973:** Endo (2008).

82   **To obtain enough:** Endo (1992).

82   **Part of the compactin molecule:** Endo (2008).

83   **Compactin looked like a promising drug:** Alfirevic et al. (2014).

83   **The bad news was:** Endo (2004).

83   **nor did the administration:** Endo (2004).

83   **One night in the spring of 1976:** Landers, P. (2006) "Stalking Cholesterol: How One Scientist Intrigued by Molds Found First Statin." *Wall Street Journal,* January 9.

83   **In just one month:** Endo (2004).

83   **Endo's success in hens:** Endo (2004).

84   **Sankyo terminated the development:** Endo (2010).

84   **Vagelos had gone to Merck from academia:** Roy Vagelos, Phone Interview, 12/17/2014.

84   **Between Brown's and Goldstein's work:** Marks (2011), pp. 1–16.

84   **Merck promptly launched a clinical trial:** Endo (2010).

85   **They also invited Endo:** Brown and Goldstein (2004).

85   **The scientists were quite surprised:** Brown et al. (1978).

85   **And if that were true, then the increased numbers:** Brown and Goldstein (2004).

85   **Sure enough, the drug increased:** Brown and Goldstein (2004); Kovanen et al. (1981).

86   **Goldstein had also seen:** Brown and Goldstein (2004).

86   **As they predicted:** Bilheimer et al. (1983); Note: homozygous FH patients do not respond to statins, they do not make functional LDL receptors and thus cannot clear LDL from the bloodstream.

86   **"Rather than replacing genes":** J. Goldstein and M. Brown, Letter to R. Vagelos, 5/5/1983; courtesy of J. Goldstein.

86   **"as fast as possible":** J. Goldstein and M. Brown, Letter to R. Vagelos, 5/5/1983; courtesy of J. Goldstein.

86   **Within a few months:** Roy Vagelos, Phone Interview, 12/17/2014; Tobert (2003).

86   **Merck's new head of Basic Research:** Edward Skolnick, Skype Interview, 12/1/2014.

86   **Goldstein and Skolnick later shared a lab:** J. Goldstein, Phone Interview, 12/1/14; Brown and Goldstein (2004).

87   **Skolnick was confident:** Skype Interview with Dr. Ed Skolnick, 12/1/2014.

87   **After two years of testing:** Havel et al. (1987); The Lovastatin Study Group III (1988).

87   **The results were better:** Scandanavian Simvastatin Survival Study Group (1994).

87   **Thanks in considerable part:** National Institutes of Health (2012), p. 10.

88   **"Without Endo, the statins":** Brown and Goldstein (2004), pp. 15–16.

CHAPTER 5

Page

89   **"The motive that will conquer":** Wells (1927), p. 44; thanks to Mike Bishop's book *How to Win a Nobel Prize* for the fitting quote.

89   **But if they happened:** Dreifus, C. (2011) "A Conversation with the Matriarch of Modern Cancer Genetics." *New York Times*, February 7; David Rowley, Phone Interview, 12/22/2014.

91　**At an early age:** Maestrejuan, A. (2005) Interview with Janet Davison Rowley. *Oral History of Human Genetics Project*, a project of UCLA and Johns Hopkins, p. 6. Available online at: ohhgp.pendari.com.

91　**When she entered medical school:** Wasserman (2000), p. 77.

92　**Although she had never:** Maestrejuan, A. (2005) Interview with Janet Davison Rowley. *Oral History of Human Genetics Project*, a project of UCLA and Johns Hopkins, p. 15.

92　**The opportunity soon came:** Chapter 2.

92　**Rowley figured that she could** Maestrejuan, A. (2005) Interview with Janet Davison Rowley. *Oral History of Human Genetics Project*, a project of UCLA and Johns Hopkins, p. 31.

92　**"I have a research project":** Dreifus, C. (2011) "A Conversation with the Matriarch of Modern Cancer Genetics." *New York Times*, February 7.

93　**Known as a *translocation*:** Maestrejuan, A. (2005) Interview with Janet Davison Rowley. *Oral History of Human Genetics Project*, a project of UCLA and Johns Hopkins, p. 40.

93　**"A favorite explanation":** Peyton Rous Nobel Lecture, "The Challenge to Man of the Neoplastic Cell." December 13, 1966. Nobelprize.org, p. 4.

93　**When she called:** Druker (2014); Maestrejuan, A. (2005) Interview with Janet Davison Rowley. *Oral History of Human Genetics Project*, a project of UCLA and Johns Hopkins, p. 46.

93　**She took pictures:** Maestrejuan, A. (2005) Interview with Janet Davison Rowley. *Oral History of Human Genetics Project*, a project of UCLA and Johns Hopkins, p. 46; David Rowley, Phone Interview, 12/22/2014.

93　**The boys teased their mom:** Druker (2014).

93　**Many years earlier:** Nowell and Hungerford (1960); Nowell (2007).

94　**This revealed that genetic information:** Maestrejuan, A. (2005) Interview with Janet Davison Rowley. *Oral History of Human Genetics Project*, a project of UCLA and Johns Hopkins, p. 46–47.

94　**The editors rejected the paper:** Maestrejuan, A. (2005) Interview with Janet Davison Rowley. *Oral History of Human Genetics Project*, a project of UCLA and Johns Hopkins, p. 47.

95　**With this additional data:** Rowley (1973).

95　**Suddenly, her part-time job** Maestrejuan, A. (2005) Interview with Janet Davison Rowley. *Oral History of Human Genetics Project*, a project of UCLA and Johns Hopkins, p. 79.

95　**Soon thereafter, Rowley identified:** Rowley et al. (1977).

95　**and another group discovered:** Zech et al. (1976).

95　**In 1910, Rockefeller Institute scientist:** Rous (1910).

96　**The critical clue was discovered:** Martin (1970); Martin (2004).

96　**If so, then one might find:** Bishop (2003), pp. 161–162.

97　**The first revealing results:** Bishop (2003).

97　***C-src* was soon found:** Stehelin, Varmus, and Bishop (1976).

97　**But *c-src* was not just:** Spector, Varmus, and Bishop (1978).

98   **A team of Dutch and British researchers:** De Klein et al. (1982).

98   **Remarkably, it turned out:** Groffen et al. (1984).

99   **But the activity of the bcr/abl:** Konopka, Watanabe, and Witte (1984).

100   **The crucial clue:** Friend et al. (1986).

101   **By zeroing in on the DNA:** Friend et al. (1986).

101   **Extensive studies of the Rb protein:** Giacinti and Giordano (2006).

101   **Rb is not the only tumor suppressor:** Vogelstein et al. (2013).

101   **Nor are Rb mutations:** Liu et al. (2004).

101   **Indeed, Rb is inactivated:** Di Fiore et al. (2013).

103   **Druker found that one particular:** Wapner (2013), pp. 140–141.

103   **Then, the first toxicology tests:** Wapner (2013), pp. 154–155.

103   **"Not over my dead body":** Wapner (2013), p. 160.

103   **His patients' prognoses were grim:** Dreifus, C. (2009) "Researcher behind the Drug Gleevec." *New York Times*, November 2.

103   **"give the drug a chance":** Wapner (2013), p. 166.

104   **In three-quarters of patients:** Druker (2003).

104   **The FDA gave the drug:** Wapner (2013), pp. 210–232.

104   **Long-term survival rates:** Hehlmann et al. (2014); Kantarjian et al. (2012).

104   **Contrary to company forecasts:** Wapner (2013), p. 265.

104   **In 2012, Lydon, Druker, and Rowley:** Press Release: "2012 Japan Prize Awarded to Trailblazers in Leukemia Research and Inventor of Highest-Performance Magnet." *The Japan Prize Foundation*. January 25, 2012. http://www.japanprize.jp/en/press_releases20120125.html.

104   **There are more than two hundred:** Bianconi et al. (2013).

105   **One major revelation:** Vogelstein et al. (2013).

105   **In 2010, Janet Rowley was:** Dizikes, C. (2013) "Janet Rowley, 1925–2013: U. of C. Scientist Made Breakthrough Cancer Discoveries." *Chicago Tribune*, December 18.

105   **She had pre-arranged:** Dizikes, C. (2013) "Janet Rowley, 1925–2013: U. of C. Scientist Made Breakthrough Cancer Discoveries." *Chicago Tribune*, December 18.

CHAPTER 6

Page

111   **"You push an ecological system":** Dietrich, B. (1996) "Tatoosh Island—Testing Nature's Limits." *Seattle Times*, September 8.

111   **On a field trip:** Robert T. Paine, Phone Interview, 4/1/2015.

111   **"Wow, this is what":** Stolzenburg (2009), p. 22.

112   **Then, with a crowbar:** Robert T. Paine, Phone Interview, 4/1/2015.

112   **Born and raised in Cambridge:** Robert T. Paine, Phone Interview, 4/1/2015.

113   **It was good discipline:** Stolzenburg (2009), p. 16.

113 **"A gang of men":** Forbush, E. H. (1927) *Birds of Massachusetts and Other New England States. Vol. 2.* Norwood, Massachusetts: Norwood Press, quoted in Paine (2011), p. 9.

113 **Young Paine spent many hours:** Paine (2011), p. 10–11.

113 **Paine appreciated how:** Robert T. Paine, Phone Interview, 4/1/2015.

114 **"We are going to stay":** Robert T. Paine, Phone Interview, 4/1/2015.

114 **"Why don't you":** Robert T. Paine, Phone Interview, 4/1/2015.

114 **Paine's first task:** Robert T. Paine, Phone Interview, 4/1/2015.

114 **Lacking a nearby ocean:** Greenberg, Herrnkind, and Coleman (2010).

114 **It was the sort of work:** Paine (1963a).

114 **And sifting large amounts of sand:** Robert T. Paine, Phone Interview, 4/1/2015.

114 **At the tip of nearby:** Paine (1963b).

115 **On top of his thesis:** Paine (1963b).

115 **"Elton (1927) suggested":** Paine (1963b), p. 65.

115 **Instead, the trio: Hairston (1989); Slobodkin (2009).**

115 **At the bottom:** Hairston, Smith, and Slobodkin (1960).

116 **That something, they believed:** Hairston, Smith, and Slobodkin (1960).

116 **It was rejected:** Hairston (1989).

116 **"The logic used is not easily refuted":** Hairston, Smith, and Slobodkin (1960), p. 421.

118 **"kick it and see" ecology:** Yong (2013), p. 287.

118 **Mussels and chitons:** Paine (1966).

119 **However, the population:** Paine (1966).

119 **As Paine continued:** Paine (1974).

119 **Paine discovered uninhabited:** Yong (2013), p. 287.

120 **Within a few months:** Paine (1974).

120 **Over a period of nine months:** Paine (1971).

121 **They observed dramatic effects:** Paine and Vadas (1969).

121 **Paine also noticed:** Paine (2010).

121 **Sea otters once ranged:** Estes and Palmisano (1974).

121 **The species gained:** Washington Department of Fish and Wildlife (2013) "Sea Otter (*Enhydra lutris*)." *Encyclopedia of Puget Sound.* Puget Sound Institute, University of Washington. http://www.eopugetsound.org/articles /sea-otter-enhydra-lutris; Doroff, A., and A. Burdin (2013) "*Enhydra lutris*." *The IUCN Red List of Threatened Species.* http://www.iucnredlist.org /details/7750/0.

121 **Some students were working:** Robert T. Paine, Phone Interview, 4/1/2015.

121 **He explained to Paine:** Stolzenburg (2009), p. 58; Eisenberg (2010), p. 60.

122 **"You want to look":** Robert T. Paine, Phone Interview, 4/1/2015.

122 **Their first hint:** Robert T. Paine, Phone Interview, 4/1/2015.

122	**But the real shock:** Stolzenburg (2009), p. 59; Estes and Palmisano (1974).
122	**They suggested that sea otters:** Stolzenburg (2009), p. 59; Estes and Palmisano (1974).
122	**But by 1978:** Duggins (1980).
123	**Paine coined a new term:** Paine (1980).
123	**These distributions suggested:** Power, Matthews, and Stewart (1985).
125	**These results not only:** Power, Matthews, and Stewart (1985).
125	**Long-term studies:** McLaren and Peterson (1994).
125	**For instance, the elimination:** Stolzenburg, W. (2008) "Ecosystems Unraveling." *Conservation Magazine.* University of Washington. http:// conservationmagazine.org/2008/07/ecosystems-unraveling.
126	**While various possible culprits:** Estes et al. (1998).
126	**They suggest that the orcas:** Estes et al. (1998); Estes, Peterson, and Steneck (2010), pp. 37–52.
127	**In another colossal:** Paine (1992).
127	**"Some animals are more":** Paine (2010), p. 25.

CHAPTER 7

Page

129	**"Africa is the continent":** Huxley (1931).
129	**The two hunters backed away:** Tony Sinclair, Personal Communication, 3/22/2015.
129	**He was awestruck:** Tony Sinclair, Phone Interview, 3/6/2015; Thomson, H. (2005) "Saving the Serengeti." *UBC Reports* 51(3). University of British Colombia. http://news.ubc.ca/ubcreports/2005/05mar03/serengeti.html.
130	**"Oh well, I am going next year":** Tony Sinclair, Phone Interview, 3/6/2015.
130	**He became good friends:** Tony Sinclair, Phone Interview, 3/6/2015.
130	**His assignment was to be Cain's assistant:** Sinclair (2012), pp. 3–4; Tony Sinclair, Phone Interview, 3/6/2015.
131	**"why it was like the way it was":** Sinclair (2012), p. 4.
131	**The American hunter Stewart Edward White:** German Oscar Baumann crossed the Serengeti in 1891 and described his encounters in *Durch Massailand zur Nilquelle* (1894).
131	**"I set out by compass":** White (1915), p. 113.
131	**"It is hard to do that country justice":** White (1915), p. 113.
131	**"Never have I seen anything":** White (1915), p. 115.
131	**"And suddenly I realized again":** White (1915), p. 115.
132	**"In her large animals":** Huxley (1931), pp. 242–243.
132	**He wrote to *The Times*:** Wheeler (2006), pp. 217–219; Sinclair (2012), p. 61.
132	**In 1937, part of this area:** The Serengeti owed its designation in part to another of Huxley's endeavors. He played an instrumental role in the

creation of UNESCO (the United Nations Educational, Scientific, and Cultural Organization) and served as its first director-general. When UNESCO convened in Stockholm in 1972 to discuss the proposal to classify outstanding natural World Heritage sites for protection, the Serengeti was right at top of the list.

133 **"the savanna of our birth":** Reid (2012), p. 1.

133 **In 1957, Bernard Grzimek:** Grzimek and Grzimek (1960); Grzimek and Grzimek (1961).

133 **For two weeks in January 1958:** Grzimek and Grzimek (1960). The Dornier 27 was a modified version of the Fieseler Storch warplane; Goebel, G. (2014), "Dornier Civil Aircraft." *AirVectors*, September 1. http://www.airvectors.net/avdojet.html.

133 **With Teutonic precision, they reported:** Grzimek and Grzimek (1960), p. 32.

133 **Altogether, they calculated 366,980:** Grzimek and Grzimek (1961), p. 136.

133 **"Were there enough plains":** Grzimek and Grzimek (1961), pp. 136–139.

133 **A survey in 1965 counted about 37,000:** Sinclair (1973a); the uncorrected estimates from Table 6 are given here.

133 **"Can a bird man do that?"** Tony Sinclair, Phone Interview, 3/6/2015.

134 **Why were there so many:** Sinclair (1977), p. 3.

134 **Sinclair joined the buffalo:** Tony Sinclair, Phone Interview, 3/6/2015.

135 **Sinclair repeated the survey:** Sinclair (1973a).

135 **In fact, by 1969:** Sinclair (1973a).

135 **To distinguish among these:** Sinclair (1974).

135 **Sinclair learned that he could:** Tony Sinclair, Phone Interview, 3/6/2015; Sinclair (1977), pp. 158–160.

135 **Sinclair examined the skulls:** Sinclair (1974).

135 **By plotting the buffalo death:** Sinclair (1973b).

135 **Field observations suggested:** Sinclair (1974), p. 193.

136 **Periodic outbreaks in Serengeti:** Sinclair (1977), pp. 252–254; Spinage (2003), p. 662.

136 **Sinclair knew that virologist:** Biggs (2010).

136 **Sinclair could therefore:** Tony Sinclair, Phone Interview, 3/6/2015.

136 **He found that while most:** Sinclair (1973b); Sinclair (1974); Sinclair (1977), p. 254; Sinclair (1979), p. 88.

136 **Sinclair examined antibody data:** Sinclair (1979), pp. 87–88.

137 **This proved that cattle:** Sinclair (1977), pp. 254–255.

138 **By 1973, the population:** Sinclair (1979), p. 91.

138 **It dawned on Sinclair:** Tony Sinclair, Phone Interview, 3/6/2015.

138 **With tensions raised:** Sinclair (2012), pp. 95–97.

139 **In addition to the vast herds:** Sinclair (2012), p. 103.

139 **"So, you are from Kenya":** Sinclair (2012), p. 103.

139 **Realizing they did not have:** Sinclair (2012), pp. 103–104.

139   **Leakey stunned Sinclair:** Sinclair (2012), p. 104; for information on foot-prints, see http://humanorigins.si.edu/evidence/behavior/laetoli-footprint-trails.

140   **A few weeks later:** Sinclair (2012), p. 105; Talbot and Stewart (1964).

140   **For instance, the number of lions:** Hanby and Bygott (1979), pp. 249–262.

140   **For example, the number of giraffe:** Grimsdell (1979), p. 358.

140   **The critical piece of the puzzle:** Norton-Griffiths (1979), pp. 310–352.

140   **The spike in wildebeest:** In grazed areas, about 700 kilograms per hectare of grass remains versus 6,200 kilograms per hectare in ungrazed areas; Sinclair et al. (2010), p. 260.

142   **"outbreak of trees":** Tony Sinclair, Phone Interview, 3/6/2015; Sinclair (2012), p. 113.

142   **"It only took a decade":** Tony Sinclair, Phone Interview, 3/6/2015.

142   **Before their increase in numbers:** Sinclair et al. (2010), pp. 255–274.

143   **These herbs in turn:** Sinclair (2003).

143   **Ecologist Sam McNaughton:** McNaughton (1979).

143   **Both the number and diversity:** McNaughton (1979).

143   **In the four years during which:** Sinclair (2012), p. 106.

143   **In contrast, the removal:** Sinclair (2003).

144   **"Without the wildebeest":** Morell, V. (2006) "The Sound of Hoofs." *Smithsonian Magazine*, June 2006. http://www.smithsonianmag.com /science-nature/the-sound-of-hoofs-119424800/?no-ist.

144   **They found a striking correlation:** Sinclair, Mduma, and Brashares (2003).

145   **For example, most smaller antelopes:** Sinclair, Mduma, and Brashares (2003), Supplemental Information, Table 2.

145   **For example, of the ten mammalian:** Sinclair, Mduma, and Brashares (2003), Supplemental Information, Table 1.

146   **And all five small-bodied:** Sinclair, Mduma, and Brashares (2003), Supplemental Information, Table 4.

146   **"the size of the prey":** Elton (1927), p. 60.

146   **They reveal a specific rule:** Sinclair (2003), p. 1732.

147   **The Grzimeks counted just sixty:** Dublin and Hamilton-Douglas (1987).

147   **When Sinclair plotted the rate:** Sinclair (2003); Sinclair et al. (2010).

149   **He discovered that a greater:** Sinclair (1977), and cited in Sinclair et al. (2010).

149   **They discovered that:** Mduma, Sinclair, and Hilborn (1999).

149   **Sinclair, Mduma, and Hilborn:** Mduma, Sinclair, and Hilborn (1999).

151   **Then, as the plains dry out:** Sinclair et al. (2010).

151   **Predation on those:** Sinclair, Mduma, and Brashares (2003).

151   **Moreover, only about 1 percent:** Fryxell, Greever, and Sinclair (1988).

151   **The combined effect:** Sinclair (2003), p. 1733.

151   **Elsewhere in Africa:** Fryxell, Greever, and Sinclair (1988).

152   **He and his wife Anne:** Tony Sinclair, Phone Interview, 3/6/2015.
152   **Thanks in large part to "Mr. Serengeti":** Morell, V. (2006) "The Sound of Hoofs." *Smithsonian Magazine*, June. http://www.smithsonianmag.com/science-nature/the-sound-of-hoofs-119424800/?no-ist.
152   **We now know about the food webs:** Morell, V. (2006) "The Sound of Hoofs." *Smithsonian Magazine*, June. http://www.smithsonianmag.com/science-nature/the-sound-of-hoofs-119424800/?no-ist.

CHAPTER 8

Page
155   **"It is failures":** Elton (1927), pp. 188–189.
155   **DO NOT DRINK:** City of Toledo, Ohio. "Urgent Water Notice." *City of Toledo website.* August 2, 2014. http://toledo.oh.gov/news/2014/08/urgent-water-notice/.
155   **The national and international news:** Wines, M. (2014) "Behind Toledo's Water Crisis, a Long-Troubled Lake Erie." *New York Times*, August 4, 2014; Mayhew, F. "Toledo Water Crisis: Half a Million People Without Safe Drinking Water as Toxins Contaminate Ohio City Supply." *The Independent*, August 3, 2014; "Division of Water Treatment." *City of Toledo website.* http://toledo.oh.gov/services/public-utilities/water-treatment/.
156   **You're glumping the pond:** Seuss (1971).
156   **Spurred by the dire condition:** "History of the Clean Water Act." *United States Environmental Protection Agency.* http://www2.epa.gov/laws-regulations/history-clean-water-act.
156   **In 1972, the United States and Canada:** Large Lakes and Rivers Forecasting Research Branch. "Detroit River-Western Lake Erie Basin Indicator Project. INDICATOR: Algal Blooms in Western Lake Erie." *United States Environmental Protection Agency.* http://www.epa.gov/med/grosseile_site/indicators/algae-blooms.html.
156   **The recovery of Lake Erie:** Egan, D. (2014) "Toxic Algae Cocktail Brews in Lake Erie." *Milwaukee Journal Sentinel*, September 13.
156   **In 2011, the lake:** Michalak et al. (2013).
157   **In a bloom, this can:** Rinta-Kanto et al. (2006); Baxa et al. (2010).
157   **The 2011 bloom may have contained:** Calculation is based on a bloom area of 5,000 square kilometers, a bloom thickness of up to 10 centimeters, and algal concentrations of 100,000 to 100 million cells per liter; alternatively, using figures of bloom size of 40,000 metric tons dry weight by Obenour et al. (2014) and 10^{11} cells per gram, the bloom would be approximately 40×10^{20} cells.
157   **When cancer spreads in a person:** "How Can Cancer Kill You?" *Cancer Research UK.* http://www.cancerresearchuk.org/about-cancer/cancers-in-general/cancer-questions/how-can-cancer-kill-you.

158   **It has plenty of company:** Michalak et al. (2013).
158   **In Cambodia, for example:** Gnanamanickam (2009).
158   **Over 30 percent of all calories:** Gnanamanickam (2009).
158   **Within ten years:** Chandler (1992).
159   **But in the mid-1970s:** Dyck and Thomas (1979).
159   **The little bugs suck the sap:** "Plantwise Technical Factsheet: Brown Planthopper (Nilaparvata Lugens)." *Plantwise Knowledge Bank.* Plantwise/ CABI. http://www.plantwise.org/KnowledgeBank/Datasheet.aspx?dsid =36301
159   **The number of bugs:** Kenmore et al. (1984).
159   **Many farmers lost nearly everything:** Sogawa (2015), p. 36.
159   **The insect was considered:** Sogawa (2015), p. 34.
159   **Indeed, insecticide treatment caused:** Kenmore et al. (1984).
159   **First, the insects had evolved:** Heinrichs (1979).
161   **A second, much more surprising:** Kenmore et al. (1984).
161   **Constant vigilance is required:** Justin Brashares in "Dangerous Catch." *National Geographic's Strange Days on Planet Earth.* (2008) PBS. Films for the Humanities & Sciences; Films Media Group; National Geographic Television & Film.
162   **The several-decades-long census:** Brashares et al. (2010).
162   **Moreover, the animals expanded:** Brashares et al. (2010).
162   **From the mid-1870s to the mid-1980s:** MacKenzie (2008).
162   **For many of the state's fishermen:** Lamontagne, N. D. (2009) "Return of the Bay Scallops." *Coastwatch.* North Carolina State University. Holiday 2009 issue. http://ncseagrant.ncsu.edu/coastwatch/previous -issues/2009–2/holiday-2009/return-of-the-bay-scallops/.
162   **The century-old fishery:** N.C. Division of Marine Fisheries. "Stock Status Report 2014." *Division of Marine Fisheries.* http://portal.ncdenr.org /web/mf/2014-stock-status-report.
163   **Peterson teamed up:** Lamontagne, N. D. (2009) "Return of the Bay Scallops." *Coastwatch.* North Carolina State University. Holiday 2009 issue. http://ncseagrant.ncsu.edu/coastwatch/previous-issues/2009–2/holiday -2009/return-of-the-bay-scallops/.
163   **They found that cownose ray populations:** Myers et al. (2007).
163   **Peterson had previously observed:** Peterson et al. (2001).
164   **Sharks eat the rays:** Myers et al. (2007).
164   **Sure enough, the researchers found:** Myers et al. (2007).
164   **A similar explanation accounts:** Brashares et al. (2010).
164   **The pesticide killed off spiders:** Kenmore et al. (1984); Schoenly et al. (1996).
165   **In healthy freshwater lakes:** Nicholls (1999).
166   **Graeme Caughley, who co-wrote:** Sinclair and Metzger (2009).
167   **"reasons for optimism":** Cannon (1928), p. 593.

CHAPTER 9

Page

169   **"Good science and good management":** Kitchell (1992), p. 4.

170   **"Fair Lakes, serene":** Brock (1985), pp. 7–8.

170   **The lake and the university:** "The History of Limnology at the University of Wisconsin." *Center for Limnology.* College of Letters and Science–University of Wisconsin website. http://limnology.wisc.edu/CFL-history .htm.

170   **Almost 90 million fish:** "Fishing Wisconsin: Planning for Panfish." *Wisconsin Department of Natural Resources.* http://dnr.wi.gov/topic/fishing/out reach/panfishplan.html.

171   **Fishermen and the general public:** Addis (1992), pp. 7–15.

171   **But Mendota is a big lake:** Addis (1992), pp. 7–15.

171   **Early adherents of the trophic:** Carpenter, Kitchell, and Hodgson (1985).

172   **They left Paul Lake untouched:** Carpenter et al. (1987).

172   **The remaining minnows:** Carpenter et al. (1987).

172   **Addis phoned Kitchell:** Addis (1992); James Kitchell, Phone Interview, 4/28/2015.

172   **Paid for by taxes:** US Fish and Wildlife Service (2000).

172   **Under the new law:** Addis (1992).

174   **Walleye are iconic fish:** James Kitchell, Phone Interview, 4/28/2015.

174   **The DNR's entire walleye:** Addis (1992).

174   **That did not go over:** James Addis, Phone Interview, 4/30/2015.

175   **So, the clubs agreed:** Johnson and Staggs (1992), pp. 353–375; James Addis, Phone Interview, 4/30/2015.

175   **By comparison, the plankton-eating cisco:** Vanni et al. (1990).

175   **Almost no walleye fry:** Johnson et al. (1992).

176   **The hot weather triggered:** Vanni et al. (1990).

176   **As a result of the cisco:** Johnson et al. (1992).

176   **After the first three years:** Lathrop et al. (2002).

176   **Moreover, the cisco remained:** Lathrop et al. (2002).

176   **Managers in countries:** Bernes et al. (2015).

177   **Around noon on January:** Milstein, M. (2015) "From Jan. 13, 1995: Return to Yellowstone: Wolves Finally Taste Freedom." *Billings Gazette*, January 13. http://billingsgazette.com/news/state-and-regional/montana/from-jan -return-to-yellowstone-wolves-finally-taste-freedom/article_69b6adf2 –57cb-57ba-b6d1–9bd9ce2a7e49.html.

177   **Babbitt peered:** Milstein, M. (2015) "From Jan. 13, 1995: Return to Yellowstone: Wolves Finally Taste Freedom." *Billings Gazette*, January 13. http:// billingsgazette.com/news/state-and-regional/montana/from-jan-return -to-yellowstone-wolves-finally-taste-freedom/article_69b6adf2–57cb -57ba-b6d1–9bd9ce2a7e49.html.

177   **Two months later:** Milstein, M. (2015) "From Jan. 13, 1995: Return to Yellowstone: Wolves Finally Taste Freedom." *Billings Gazette*, January 13. http://billingsgazette.com/news/state-and-regional/montana/from-jan -return-to-yellowstone-wolves-finally-taste-freedom/article_69b6adf2 –57cb-57ba-b6d1–9bd9ce2a7e49.html.

177   **For the species:** Smith (2005), p. 8.

177   **The Yellowstone Wolf:** Bangs (2005), p. 4.

178   **Elk populations irrupted:** Ripple, Rooney, and Beschta (2010).

179   **The result was the Environmental Impact Statement:** US Fish and Wildlife Service (1994).

179   **For livestock, the EIS:** White et al. (2005).

179   **Ten years after the release:** White et al. (2005).

179   **Confirmed kills of livestock:** http://www.livescience.com/49897-idaho -board-spending-4600-on-every-wolf-killed.html?cmpid=559188.

179   **The impact on elk:** White et al. (2005).

180   **It had been noted:** Ripple and Beschta (2005).

180   **To investigate further:** "In the Valley of the Wolves: Reintroduction of the Wolves." *Nature*. PBS. July 13, 2011. http://www.pbs.org/wnet/nature /in-the-valley-of-the-wolves-reintroduction-of-the-wolves/213; Ripple and Larsen (2000).

180   **They knew that elk:** Eisenberg, Seager, and Hibbs (2013).

180   **"the re-establishment":** Ripple and Larsen (2000), p. 367.

181   **The willows he did see:** Eisenberg (2010), p. 97.

181   **He examined the cottonwood:** Beschta (2005).

181   **Browsing appeared to be reduced:** Ripple and Beschta (2007); Ripple and Beschta (2012).

181   **After the wolves' return:** Ripple and Beschta (2012).

182   **Long-term studies:** Berger, Gese, and Berger (2008).

182   **Similarly, researchers in Yellowstone:** Marshall, Hobbs, and Cooper (2013).

CHAPTER 10

Page

183   **Politics we shall:** Kellaway, K. (2010) "How the Observer Brought the WWF into Being." *The Guardian*, November 6. http://www.theguardian .com/environment/2010/nov/07/wwf-world-wildlife-fund-huxley.

184   **He estimated populations of 14,000 buffalo:** Dunham, K. (2005) "Mammals: Re-introductions of Large Mammals in Gorongosa National Park, Mozambique." *Re-Introduction* 24: 10–11.

184   **The Gorongosa area lion population:** Wentzel (1964), pp. 230–231; "Gorongosa's Lions." *Gorongosa National Park Website*. http://www .gorongosa.org/explore-park/wildlife/lions-gorongosa.

184   **Over the course of forty days:** Wentzel (1964), pp. 230–231; "Gorongo-sa's Lions." *Gorongosa National Park Website.* http://www.gorongosa.org /explore-park/wildlife/lions-gorongosa; Cumming et al. (1994).

186   **"A Dream Becomes a Nightmare":** Dutton (1994).

186   **Over the course of fifteen:** Gourevitch (2009).

186   **In December 1981:** Morley and Convery (2014), p. 133.

186   **Even after a peace accord:** Morley and Convery (2014), p. 134.

186   **The small, seasoned team:** Morley and Convery (2014), p. 134.

188   **He had endowed a Center:** Hanes, S. (2007). "Greg Carr's Big Gamble." *Smithsonian.com.* May 2007. http://www.smithsonianmag.com/people -places/greg-carrs-big-gamble-153081070/?no-ist.

188   **But Carr was looking:** Greg Carr, Interview, Gorongosa National Park, 6/22/2015.

188   **"You should go talk to him":** Greg Carr, Phone Interview, 5/19/2015.

188   **"Come to Mozambique":** Greg Carr, Phone Interview, 5/19/2015.

188   **"adopt a project":** Greg Carr, Phone Interview, 5/19/2015.

188   **Seeing both the beauty:** Greg Carr, Interview, Gorongosa National Park, 6/22/2015.

189   **So he began devouring:** Greg Carr, Interview, Gorongosa National Park, 6/22/2015.

189   **They then made:** Greg Carr, Phone Interview, 5/19/2015.

189   **"This is a spectacular park":** "Gorongosa: Restoring Mozambique's National Treasure." *BioInteractive.* HHMI. https://www.hhmi.org/bio interactive/gorongosa-restoring-mozambiques-national-treasure.

189   **When Carr went back:** Shacochis, B. (2009) "Saving Gorongosa." *Out-side,* June 20. http://www.outsideonline.com/1885571/saving-gorongosa.

189   **"Don't bother":** Hanes, S. (2007). "Greg Carr's Big Gamble." *Smithsonian. com.* May 2007. http://www.smithsonianmag.com/people placs/greg -carrs-big-gamble-153081070/?no-ist.

190   **The results were mixed:** Dunham (2004).

190   **Unable to move:** Greg Carr, Personal Communication, 6/2015.

191   **The first batch:** "Timeline." *Gorongosa National Park Website.* http:// www.gorongosa.org/our-story/timeline.

191   **Gorongosa had been home:** "Zebra and Eland Relocated to Goron-gosa!" *Gorongosa Blog.* September 17, 2013. http://www.gorongosa.org /blog/park-news/zebra-and-eland-relocated-gorongosa.

191   **Carr's scientific team:** "Zebra and Eland Relocated to Gorongosa!" *Gorongosa Blog.* September 17, 2013. http://www.gorongosa.org/blog /park-news/zebra-and-eland-relocated-gorongosa.

192   **One hundred-eighty wildebeest:** "Zebra and Eland Relocated to Goron-gosa!" *Gorongosa Blog.* September 17, 2013. http://www.gorongosa.org/blog /park-news/zebra-and-eland-relocated-gorongosa; "Gorongosa: Restoring Mozambique's National Treasure." *BioInteractive.* HHMI. https://www

.hhmi.org/biointeractive/gorongosa-restoring-mozambiques-national -treasure.

192　**I went to Gorongosa to see for myself:** The author stayed in and toured Gorongosa National Park June 17–26, 2015.

195　**In creating jobs for:** Greg Carr, Interview, Gorongosa National Park, 6/22/2015.

195　**The rangers also have:** "Preventing Human-Elephant Conflict." *Gorongosa National Park: Ranger Diary.* http://www.gorongosa.org/our-story /conservation/preventing-human-elephant-conflict.

195　**While fewer than 1,000 tourists:** Shacochis, B. (2009) "Saving Gorongosa." *Outside.* 20 Jun 2009. http://www.outsideonline.com/1885571 /saving-gorongosa.

195　**In 2009, a mobile health clinic:** "Timeline." *Gorongosa National Park Website.* http://www.gorongosa.org/our-story/timeline.

196　**To help curb malaria:** Gourevitch (2009).

197　**Over a span of twelve days:** October 24–November 4, 2014; "Aerial Wildlife Count."*Gorongosa Blog.* December 14, 2014. http://www .gorongosa.org/blog/park-news/aerial-wildlife-count.

198　**nineteen species of herbivores in all:** Stalmans, Peel, and Massad (2014).

198　**"You give nature":** Gourevitch (2009), p. 100.

199　**For example, the northern elephant seal:** Roman et al. (2015); "Basic Facts about American Alligators." *Defenders of Wildlife.* http://www .defenders.org/american-alligator/basic-facts.

199　**"Focus on law enforcement":** Greg Carr, Phone Interview, 5/19/2015.

199　**While they have rebounded:** Paola Bouley, Personal Communication, 6/22/2015.

199　**Stalmans has estimated:** Stalmans, Peel, and Massad (2014).

200　**Biologists have estimated:** Naskrecki, P. (2015) "Charting the Map of Gorongosa's Life." *Gorongosa Field Notes: E.O. Wilson Biodiversity Laboratory at Gorongosa National Park.* E.O. Wilson Biodiversity Foundation. http://eowilsonfoundation.org/gorongosa-field-notes-e-o-wilson-bio diversity-laboratory-at-gorongosa-national-park.

200　**Thanks to our extraordinary guide:** Civets are not cats, they and genets belong to a family called Viverridae.

AFTERWORD

Page

203　**"Humans are certainly":** Paine (2010), p. 35.

203　**Frequently, when certain species:** Pope Francis, July 18, 2015. *Encyclical Letter Laudato Si of the Holy Father Francis On Care For Our Common Home.* Rome: Vatican Press, p. 27. http://w2.vatican.va/content/dam/francesco

/pdf/encyclicals/documents/papa-francesco_20150524_enciclica-laudato
-si_en.pdf.

203   **"to address every person":** Pope Francis, July 18, 2015. *Encyclical Letter Laudato Si of the Holy Father Francis On Care For Our Common Home.* Rome: Vatican Press, p.4. http://w2.vatican.va/content/dam/francesco/pdf/encyclicals/documents/papa-francesco_20150524_enciclica-laudato-si_en.pdf.

204   **[W]e need only take a frank look:** Pope Francis, July 18, 2015. *Encyclical Letter Laudato Si of the Holy Father Francis On Care For Our Common Home.* Rome: Vatican Press, p. 44. http://w2.vatican.va/content/dam/francesco/pdf/encyclicals/documents/papa-francesco_20150524_enciclica-laudato-si_en.pdf.

204   **"ecological conversion":** Pope Francis, July 18, 2015. *Encyclical Letter Laudato Si of the Holy Father Francis On Care For Our Common Home.* Rome: Vatican Press, p. 159. http://w2.vatican.va/content/dam/francesco/pdf/encyclicals/documents/papa-francesco_20150524_enciclica-laudato-si_en.pdf.

205   **Just fifty years ago:** Henderson (2011).

205   **"Social considerations make it":** Dubos (1965), p. 379.

205   **The dismayed director:** Henderson (2011); Tucker (2001), p. 59.

205   **In practice, finding and reaching:** Foege (2011), p. 53.

206   **Ring vaccination removed:** Foege (2011), pp. 15, 58.

206   **Even though just 15:** Tucker (2001), p. 77.

207   **This experience in West Africa:** Foege (2011), p. 100.

207   **The WHO came up with the strategy:** Henderson (2011).

207   **The search teams quickly discovered:** Foege (2011), pp. 114–115.

207   **With one hundred new outbreaks:** Foege (2011), pp. 164–170.

208   **Bangladesh and then Ethiopia:** Henderson (2011).

208   **The disease was officially:** Henderson (2011).

208   **"It was a victory achieved":** Henderson (2011).

208   **After decades of effort:** Roeder (2011).

208   **These workers were then able:** Blanding, M. (2011) "Contagion." *Tufts Magazine.* Tufts University. Fall 2011. http://www.tufts.edu/alumni/magazine/fall2011/features/contagion.html.

209   **In his conclusion:** Foege (2011), pp. 188–192.

209   **"This is a cause-and-effect":** Foege (2011), p. 188.

210   **"If we are truly to develop an ecology":** Pope Francis, July 18, 2015. *Encyclical Letter Laudato Si of the Holy Father Francis On Care For Our Common Home.* Rome: Vatican Press, pp. 45–46. http://w2.vatican.va/content/dam/francesco/pdf/encyclicals/documents/papa-francesco_20150524_enciclica-laudato-si_en.pdf.

210   **Henderson reached for the microphone:** Hopkins (1989), p. 134.

211   **The government also banned:** Food and Agriculture Organization of the United Nations (1998) "IPM-trained Farmers in Indonesia Escape Pest Outbreaks." Rome: Food and Agriculture Organization of the United

Nations. http://www.fao.org/english/newsroom/highlights/1998/981104-e .htm; However, after changes in the Indonesian government, pesticide use has come resurged, and so have planthopper plagues: Thornburn (2015).

211   "The trouble with being an optimist": Foege (2011), pp. 188–192.

211   "Choose optimism": "The Interview—Greg Carr, Gonongosa N. P." *Passage to Africa*. http://www.passagetoafrica.com/articles/172-the-interview-greg -carr-gorongosa-n-p.

211   "What is the goal of our work": Pope Francis, July 18, 2015. *Encyclical Letter Laudato Si of the Holy Father Francis On Care For Our Common Home*. Rome: Vatican Press, p. 119. http://w2.vatican.va/content/dam/francesco /pdf/encyclicals/documents/papa-francesco_20150524_enciclica-laudato -si_en.pdf.

212   **Ali Maow Maalin:** Tucker (2001), p. 116.

212   "It looked like the shot hurt": Donnelly, J. (2006) "Polio: A Fight in a Lawless Land." *Boston Globe*, February 27, 2006. http://www.boston.com /yourlife/health/diseases/articles/2006/02/27/polio_a_fight_in_a_lawless _land.

212   **Checkpoints were also set up:** Tucker (2001), pp. 116–118.

213   "I tell the community": "Somalia Is Again Polio-Free." World Health Organization Media Centre. WHO podcast. March 28, 2008. http://www .who.int/mediacentre/multimedia/podcasts/2008/transcript_30/en/.

213   "Somalia was the last country with smallpox": "War-Torn Somalia Eradicates Polio." *BBC News*, March 25, 2008. http://news.bbc.co.uk/2/hi /africa/7312603.stm.

BIBLIOGRAPHY

Abelson, P. H., E. T. Bolton, and E. Aldous (1952) "Utilization of Carbon Dioxide in the Synthesis of Proteins by *Escherichia coli*. II." *Journal of Biological Chemistry* 198: 173–178.

Addis, J. T. (1992) "Policy and Practice in UW-WDNR Collaborative Programs." In J. F. Kitchell (ed.), *Food Web Management: A Case Study of Lake Mendota.* New York: Springer-Verlag: 7–16.

Alfirevic, A., D. Neely, J. Armitage, H. Chinoy, et al. (2014) "Phenotype Standardization for Statin-Induced Myotoxicity." *Clinical Pharmacology & Therapeutics* 96(4): 470–476.

Anderson, N. L., and N. G. Anderson (2002) "The Human Plasma Proteome: History, Character, and Diagnostic Prospects." *Molecular and Cellular Proteomics* 1: 845–867.

Anker, P. (2001) *Imperial Ecology: Environmental Order in the British Empire, 1895–1945.* Cambridge, Massachusetts: Harvard University Press.

Bangs, E. (2005) "How Did Wolves Get Back to Yellowstone?" *Yellowstone Science* 13: 4.

Barker, R. (2005) *Blockade Busters: Cheating Hitler's Reich of Vital War Supplies.* South Yorkshire, England: Pen & Sword Books.

Baxa, D. V., T. Kurobe, K. A. Ger, P. W. Lehman, and S. J. Teh (2010) "Estimating the Abundance of Toxic Microcystis in the San Francisco Estuary Using Quantitative Real-Time PCR." *Harmful Algae* 9: 342–349.

Benison, S., A. C. Barger, and E. L. Wolfe (1987) *Walter B. Cannon: The Life and Times of a Young Scientist.* Cambridge, Massachusetts: Harvard University Press.

Berger, K. M., E. M. Gese, and J. Berger (2008) "Indirect Effects and Traditional Trophic Cascades: A Test Involving Wolves, Coyotes, and Pronghorn." *Ecology* 89(3): 818–828.

Bernes, C., S. R. Carpenter, A. Gardmark, P. Larsson, et al. (2015). "What Is the Influence of a Reduction of Planktivorous and Benthivorous Fish on Water Quality in Temperate Eutrophic Lakes? A Systematic Review." *Environmental Evidence* 4:7: 1–28.

Beschta, R. L. (2005) "Reduced Cottonwood Recruitment Following Extirpation of Wolves in Yellowstone's Northern Range." *Ecology* 86: 391–403.

Bianconi, E., A. Piovesan, F. Facchin, A. Beraudi, et al. (2013) "An Estimation of the Number of Cells in the Human Body." *Annals of Human Biology* Early Online 40: 1–11.

Biggs, P. M. (2010) "Walter Plowright. 20 July 1923–20 February 2010." *Biographical Memoirs of Fellows of the Royal Society* 56: 341–358.

Bilheimer, D. W., S. M. Grundy, M. S. Brown, and J. L. Goldstein. (1983) "Mevinolin and Colestipol Stimulate Receptor-Mediated Clearance of Low Density Lipoprotein from Plasma in Familial Hypercholesterolemia Heterozygotes." *Proceedings of the National Academy of Sciences USA* 80(13): 4124–4128.

Binney, G. (1926) *With Seaplane and Sledge in the Arctic*. New York: George H. Doran.

Bishop, J. M. (2003) *How to Win the Nobel Prize: An Unexpected Life in Science*. London: Harvard University Press.

Brashares, J. S., L. R. Prugh, C. J. Stoner, and C. W. Epps (2010) "Ecological and Conservation Implications of Mesopredator Release." In J. Terborgh and J. A. Estes (eds.), *Trophic Cascades: Predators, Prey, and the Changing Dynamics of Nature*. Washington, DC: Island Press: 221–240.

Brock, T. D. (1985) *A Eutrophic Lake: Lake Mendota, Wisconsin*. New York: Springer-Verlag.

Brown, M. S., and J. L. Goldstein (1975) "Regulation of the Activity of the Low Density Lipoprotein Receptor in Human Fibroblasts." *Cell* 6: 307–316.

——— (1981) "Lowering Plasma Cholesterol by Raising LDL Receptors." *New England Journal of Medicine* 305(9): 515–517.

——— (1993) "A Receptor-Mediated Pathway for Cholesterol Homeostasis." In T. Frängsmyr and J. Lindsten (eds.), *Nobel Lectures, Physiology of Medicine 1981–1990*. Singapore: World Scientific: 284–324. Available at nobelprize.org.

——— (2004) "A Tribute to Akira Endo, Discoverer of a 'Penicillin' for Cholesterol." *Atherosclerosis Supplements* 5: 13–16.

Brown, M. S., S. E. Dana, and J. L. Goldstein (1973) "Regulation of 3-Hydroxy-3-Methylglutaryl Coenzyme A Reductase Activity in Human Fibroblasts by Lipoproteins." *Proceedings of the National Academy of Sciences USA* 70(7): 2162–2166.

——— (1974) "Regulation of 3-Hydroxy-3-Methylglutaryl Coenzyme A Reductase Activity in Cultured Human Fibroblasts." *Journal of Biological Chemistry* 249: 789–796.

Brown, M. S., J. R. Faust, J. L. Goldstein, I. Kaneko, and A. Endo (1978) "Induction of 3-Hydroxy-3-Methylglutaryl Coenzyme A Reductase Activity in Human

Fibroblasts Incubated with Compactin (ML-236B), a Competitive Inhibitor of the Reductase." *Journal of Biological Chemistry* 253: 1121–1128.

Cannon, W. B. (1873–1945) Walter Bradford Cannon papers, 1873–1945, 1972–1974 (inclusive), 1881–1945 (bulk). H MS c40. Harvard Medical Library, Francis A. Countway Library of Medicine, Boston.

——— (1898) "The Movements of the Stomach Studied by Means of the Röntgen Rays." *American Journal of Physiology* 1: 359–382.

——— (1909) "The Influence of Emotional States on the Functions of the Alimentary Canal." *American Journal of Medical Sciences* 137: 480–487.

——— (1911a) *The Mechanical Factors of Digestion*. London: Edward Arnold.

——— (1911b) "The Stimulation of Adrenal Secretion by Emotional Excitement." *Proceedings of the American Philosophical Society* 50(199): 226–227.

——— (1914) "The Emergency Function of the Adrenal Medulla in Pain and the Major Emotions." *American Journal of Physiology* 33: 356–372.

——— (1927) *Bodily Changes in Pain, Hunger, Fear and Rage: An Account of Recent Researches into the Function of Emotional Excitement*. New York and London: D. Appleton and Company.

——— (1928) "Reasons for Optimism in the Care of the Sick." *New England Journal of Medicine* 199(13): 593–597.

——— (1929) "Organization for Physiological Homeostasis." *Physiological Reviews* 9(3): 399–431.

——— (1963) *The Wisdom of the Body*. Revised and enlarged ed. New York: W. W. Norton & Company.

——— (1972) *The Life and Contributions of Walter Bradford Cannon*. Edited by C. McC. Brooks, K. Koizumi, and J. O. Pinkston. New York: Downstate Medical Center, State University of New York.

Cannon, W. B., and D. de la Paz (1911) "Emotional Stimulation of Adrenal Secretion." *American Journal of Physiology* 28: 64–70.

Carpenter, S. R., J. F. Kitchell, and J. R. Hodgson (1985) "Cascading Trophic Interactions and Lake Productivity." *BioScience* 35(10): 634–639.

Carpenter, S. R., J. F. Kitchell, J. R. Hodgson, P. A. Cochran, et al. (1987) "Regulation of Lake Primary Productivity by Food Web Structure." *Ecology* 68(6): 1863–1876.

Carroll, S. B. (2013) "Brave Genius: a Scientist, a Philosopher, and Their Daring Adventures from the French Resistance to the Nobel Prize." New York: Crown.

Chandler, R. F. Jr. (1992) *An Adventure in Applied Science: A History of the International Rice Research Institute*. Los Baños, Laguna, Philippines: International Rice Research Institute.

Cumming, D.H.M, C. Mackie, S. Magane, and R. D. Taylor (1994) *Aerial Census of Large Herbivores in the Gorongosa National Park and the Marromeu Area of the Zambezi Delta in Mozambique*. Unpublished report, IUCN ROSA, Harare, 10 pp., cited in K. M. Dunham (2004) "Aerial Survey of Large Herbivores in

Gorongosa National Park, Mozambique: 2004." *A Report for The Gregory C. Carr Foundation*. http://www.carrfoundation.org.

Darwin, C. (1872) *The Origin of Species by Means of Natural Selection, or The Preservation of Favoured Races in the Struggle for Life*. Sixth ed. London: John Murray.

Debré, P. (1996) *Jacques Monod*. Paris: Flammarion.

De Klein, A., A. G. van Kessel, G. Grosveld, C. R. Bartram, et al. (1982) "A Cellular Oncogene Is Translocated to the Philadelphia Chromosome in Chronic Myelocytic Leukaemia." *Nature* 300: 765–767.

Di Fiore, R., A. D'Anneo, G. Tesoriere, and R. Vento (2013) "RB1 in Cancer: Different Mechanisms of RB1 Inactivation and Alterations of pRb Pathway in Tumorigenesis." *Journal of Cellular Physiology* 228:1676–1687.

Drach, P., and J. Monod (1935) "Rapport préliminaire sur les observations d'histoire naturelle faites pendant la camoagne de la 'Pourquios-Pas?' au Groenland." *Annales Hydrographiques* 2: 3–11.

Druker, B. J. (2003) "David A. Karnofsky Award Lecture. Imatinib as a Paradigm of Targeted Therapies." *Journal of Clinical Oncology* 21(23 suppl): 239s–245s.

——— (2014) "Janet Rowley (1925–2013)." *Nature* 505: 484.

Dublin, H. T., and I. Douglas-Hamilton (1987) "Status and Trends of Elephants in the Serengeti-Mara Ecosystem." *African Journal of Ecology* 25: 19–33.

Dubos, R. J. (1965) *Man Adapting*. New Haven, Connecticut: Yale University Press.

Duggins, D. O. (1980) "Kelp Beds and Sea Otters: An Experimental Approach." *Ecology* 61(3): 447–453.

Dulvy, N. K., S. L. Fowler, J. A. Musick, R. D. Cavanagh, et al. (2014). "Extinction Risk and Conservation of the World's Sharks and Rays." *eLife* 3: e00590.

Dunham, K. M. (2004) "Aerial Survey of Large Herbivores in Gorongosa National Park, Mozambique: 2004." Report for the Gregory C. Carr Foundation. http://www.carrfoundation.org.

Dutton, P. (1994) "A Dream Becomes a Nightmare." *African Wildlife* 48(6): 6–14.

Dyck, V. A., and B. Thomas (1979) "The Brown Planthopper Problem." In *Brown Planthopper: Threat to Rice Production in Asia*. Los Baños, Philippines: International Rice Research Institute: 3–17.

Eisenberg, C. (2010) *The Wolf's Tooth: Keystone Predators, Trophic Cascades, and Biodiversity*. Washington, DC: Island Press.

Eisenberg, C., S. T. Seager, and D. E. Hibbs (2013) "Wolf, Elk, and Aspen Food Web Relationships: Context and Complexity." *Forest Ecology and Management* 299: 70–80.

Elton, C. S. (1924) "Periodic Fluctuations in the Numbers of Animals: Their Causes and Effects." *British Journal of Experimental Biology* 2: 119–163.

——— (1927) *Animal Ecology*. New York: Macmillan.

——— (1983) "The Oxford University Expedition to Spitsbergen in 1921: An Account, Done in 1978–1983." Norsk Polarinstitutt Bibliotek, Norsk Polarinstitutt, Oslo. http://brage.bibsys.no/xmlui/handle/11250/218913.

Elton, C. S., and M. Nicholson (1942) "The Ten-Year Cycle in Numbers of the Lynx in Canada." *Journal of Animal Ecology* 11(2): 215–244.

Endo, A. (1992) "The Discovery and Development of HMG-CoA Reductase Inhibitors." *Journal of Lipid Research* 33: 1569–1582.

—— (2004) "The Origin of the Statins." *Atherosclerosis Supplements* 5: 125–130.

—— (2008) "A Gift From Nature: The Birth of the Statins." *Nature Medicine* 14(10): 1050–1025.

—— (2010) "A Historical Perspective on the Discovery of Statins." *Proceedings of the Japan Academy, Series B* 86: 484–498.

Estes, J. A., and J. F. Palmisano (1974) "Sea Otters: Their Role in Structuring Nearshore Communities." *Science* 185: 1058–1060.

Estes, J. A., C. H. Peterson, and R. S. Steneck (2010) "Some Effects of Apex Predators in Higher-Latitude Coastal Oceans." In J. Terborgh and J. A. Estes (eds.), *Trophic Cascades: Predators, Prey, and the Changing Dynamics of Nature*. Washington, DC: Island Press: 37–53.

Estes, J. A., M. T. Tinker, T. M. Williams, and D. F. Doak (1998) "Killer Whale Predation on Sea Otters Linking Oceanic and Nearshore Ecosystems." *Science* 282: 473–476.

Everly, G. S., and J. M. Lating (2013) *A Clinical Guide to the Treatment of the Human Stress Response*. New York: Springer.

Finger, S. (1994) *Origins of Neuroscience: A History of Explorations into Brain Function*. New York: Oxford University Press.

Fleming, D. (1984) "Walter B. Cannon and Homeostasis." *Social Research* 51(3): 609–640.

Foege, W. H. (2011) *House on Fire: The Fight to Eradicate Smallpox*. Berkeley: University of California Press.

Friend, S. H., R. Bernards, S. Rogelj, R. A. Weinberg, et al. (1986) "A Human DNA Segment with Properties of the Gene that Predisposes to Retinoblastoma and Osteosarcoma." *Nature* 323: 643–646.

Fryxell, J. M., J. Greever, and A.R.E. Sinclair (1988) "Why Are Migratory Ungulates So Abundant?" *American Naturalist* 131(6): 781–798.

Giacinti, C., and A. Giordano (2006) "RB and Cell Cycle Progression." *Oncogene* 25: 5220–5227.

Gnanamanickam, S. S. (2009) "Rice and Its Importance to Human Life." In *Biological Control of Rice Diseases*. Dordrecht: Springer: 1–11.

Goldstein, D. S. (2010) "Adrenal Responses to Stress." *Cellular and Molecular Neurobiology* 30(8): 1433–1440.

Goldstein, D. S., and M. S. Brown (1973) "Familial Hypercholesterolemia: Identification of a Defect in the Regulation of 3-Hydroxy-3-Methylglutaryl Coenzyme A Reductase Activity Associated with Overproduction of Cholesterol." *Proceedings of the National Academy of Sciences USA* 70(10): 2804–2808.

—— (1974) "Familial Hypercholesterolemia: Defective Binding of Lipoproteins to Cultured Fibroblasts Associated with Impaired Regulation of

3-Hydroxy-3-Methylglutaryl Coenzyme A Reductase Activity." *Proceedings of the National Academy of Sciences USA* 71(3): 788–792.

———— (2003) "Cholesterol: A Century of Research." *HHMI Bulletin*, September: 10–19. Available at http://www4.utsouthwestern.edu/moleculargenetics/pdf /msb_cur_res/2003%20HHMI%20Bulletin%20Goldstein%2018.htm.

Gordon, S. (1922) *Amid Snowy Wastes: Wild Life on the Spitsbergen Archipelago.* New York: Cassell and Company.

Gould, R. G., C. B. Taylor, J. S. Hagerman, I. Warner, and D. J. Campbell (1953) "Cholesterol Metabolism: I. Effect of Dietary Cholesterol on the Synthesis of Cholesterol in Dog Tissue in Vitro." *Journal of Biological Chemistry* 201: 519–528.

Gourevitch, P. (2009) "The Monkey and the Fish." *New Yorker*, December 21: 99–111.

Greenberg, M. J., W. F. Herrnkind, and F. C. Coleman (2010) "Evolution of the Florida State University Coastal and Marine Laboratory." *Gulf of Mexico Science* 1–2: 149–163.

Grimsdell, J.J.R. (1979) "Changes in Populations of Resident Ungulates." In A.R.E. Sinclair and M. Norton-Griffiths (eds.), *Serengeti, Dynamics of an Ecosystem.* Chicago: University of Chicago Press: 353–359.

Groffen, J., J. R. Stephenson, N. Heisterkamp, A. de Klein, et al. (1984) "Philadelphia Chromosomal Breakpoints Are Clustered within a Limited Region, bcr, on Chromosome 22." *Cell* 36: 93–99.

Grzimek, B., and M. Grzimek (1961) *Serengeti Shall Not Die.* New York: E. P. Dutton & Co.

Grzimek, M., and B. Grzimek (1960) "Census of Plains Animals in the Serengeti National Park, Tanganyika." *Journal of Wildlife Management* 24(1): 27–37.

Hairston, N. G. (1989) *Ecological Experiments: Purpose, Design and Execution.* Cambridge: Cambridge University Press.

Hairston, N. G., F. E. Smith, and L. B. Slobodkin (1960) "Community Structure, Population Control, and Competition." *American Naturalist* 94(879): 421–425.

Hanby, J. P., and J. D. Bygott (1979) "Population Changes in Lions and Other Predators." In A.R.E. Sinclair and M. Norton-Griffiths (eds.), *Serengeti, Dynamics of an Ecosystem.* Chicago: University of Chicago Press: 249–262.

Havel, R. J., D. B Hunninghake, D. R. Illingworth, R. S. Lees, et al. (1987) "Lovastatin (Mevinolin) in the Treatment of Heterozygous Familial Hypercholesterolemia: A Multicenter Study." *Annals of Internal Medicine* 107(5): 609–615.

Hehlmann, R., M. C. Müller, M. Lauseker, B. Hanfstein, et al. (2014) "Deep Molecular Response Is Reached by the Majority of Patients Treated with Imatinib, Predicts Survival, and Is Achieved More Quickly by Optimized High-Dose Imatinib: Results from the Randomized CML-Study IV." *Journal of Clinical Oncology* 32: 415–423.

Heinrichs, E. A. (1979) "Chemical Control of the Brown Planthopper." In *Brown Planthopper: Threat to Rice Production in Asia.* Los Baños, Philippines: International Rice Research Institute: 145–167.

Heisterkamp, N., and J. Groffen (2002) "Philadelphia-Positive Leukemia: A Personal Perspective." *Oncogene* 21: 8536–8540.

Henderson, D. A. (2011) "On the Eradication of Smallpox and the Beginning of a Public Health Career." *Public Health Reviews* 33: 19–29.

Hogness, D., M. Cohn, and J. Monod (1955) "Studies on the Induced Synthesis of β-galactosidase in *Escherichia coli*: The Kinetics and Mechanism of Sulfur Incorporation." *Biochimica et Biophysica Acta* 16: 99–116.

Hopkins, J. W. (1989) *The Eradication of Smallpox: Organizational Learning and Innovation in International Health*. Boulder, Colorado: Westview Press.

Huxley, J. (1931) *Africa View*. London: Chatto & Windus.

Jacob, F. (1973) *The Logic of Life: A History of Heredity*. New York: Pantheon Books.

———— (1988) *The Statue Within*. New York: Basic Books.

Jacob F., and J. Monod (1963) "Elements of Regulatory Circuits in Bacteria." In R.J.C. Harriss (ed.), *Biological Organization at the Cellular and Supercellular Level; A Symposium Held at Varenna, 24–27 September, 1962, under the Auspices of UNESCO*. London and New York: Academic Press: 1–24.

Johnson, B. M., and M. D. Staggs. (1992) "The Fishery." In J. F. Kitchell (ed.), *Food Web Management: A Case Study of Lake Mendota*. New York: Springer-Verlag: 353–376.

Johnson, B. M., S. J. Gilbert, R.S.S. Stewart, L. G. Rudstam, et al. (1992) "Piscivores and Their Prey." In J. F. Kitchell (ed.), *Food Web Management: A Case Study of Lake Mendota*. New York: Springer-Verlag: 319–352.

Judson, H. F. (1979) *The Eighth Day of Creation: The Makers of the Revolution in Biology*. New York: Simon and Schuster.

Kantarjian, H., S. O'Brien, E. Jabbour, G. Garcia-Manero, et al. (2012) "Improved Survival in Chronic Myeloid Leukemia Since the Introduction of Imatinib Therapy: A Single-Institution Historical Experience." *Blood* 119(9): 1981–1987.

Kenmore, P. E., F. O. Carino, C. A. Perez, V. A. Dyck, and A. P. Gutierrez (1984) "Population Regulation of the Rice Brown Planthopper (*Nilaparvata lugens* Stål) within Rice Fields in the Philippines." *Journal of Plant Protection in the Tropics* 1: 19–37.

Keys, A. (1990) "Recollections of Pioneers in Nutrition: From Starvation to Cholesterol." *Journal of the American College of Nutrition* 9(4): 288–291.

Keys, A., H. L. Taylor, H. Blackburn, J. Brozek, et al. (1963) "Coronary Heart Disease among Minnesota Business and Professional Men Followed Fifteen Years." *Circulation* 28: 381–395.

Kitchell, J. F. (1992) *Food Web Management: A Case Study of Lake Mendota*. New York: Springer-Verlag.

Konopka, J. B., S. M. Watanabe, and O. N. Witte (1984) "An Alteration of the Human *c-abl* Protein in K562 Leukemia Cells Unmasks Associated Tyrosine Kinase Activity." *Cell* 37: 1035–1042.

Kovanen, P. T., D. W. Bilheimer, J. L. Goldstein, J. J. Jaramillo, and M. S. Brown (1981) "Regulatory Role for Hepatic Low Density Lipoprotein Receptors in vivo in the Dog." *Proceedings of the National Academy of Sciences USA* 78(2): 1194–1198.

Lathrop, R. C., B. M. Johnson, T. B. Johnson, M. T. Vogelsang, et al. (2002) "Stocking Piscivores to Improve Fishing and Water Clarity: A Synthesis of the Lake Mendota Biomanipulation Project." *Freshwater Biology* 47: 2410–2424.

Ledford, H. (2011) "Translational Research: 4 Ways to Fix the Clinical Trial." *Nature* 477: 526–528.

Lindström, J., E. Ranta, H. Kokko, P. Lundberg, and V. Kaitala (2001) "From Arctic Lemmings to Adaptive Dynamics: Charles Elton's Legacy in Population Ecology." *Biological Reviews of the Cambridge Philosophical Society* 76(1): 129–158.

Liu, H., B. Dibling, B. Spike, A. Dirlam, and K. Macleod (2004) "New Roles for the RB Tumor Suppressor Protein." *Current Opinion in Genetics & Development* 14(1): 55–64.

Longstaff, T. (1950) *This My Voyage*. New York: Carles Scribner's Sons.

Lovastatin Study Group III (1988) "A Multicenter Comparison of Lovastatin and Cholestyramine Therapy for Severe Primary Hypercholesterolemia." *Journal of the American Medical Association* 260(3): 359–366.

Lwoff, A. (2003) "Jacques Lucien Monod." In A. Ullmann (ed.), *Origins of Molecular Biology: a Tribute to Jacques Monod*. Revised ed. Washington, DC: ASM Press: 1–23.

Lydon, N. B., and B. J. Druker (2004) "Lessons Learned from the Development of Imatinib." *Leukemia Research* 28S1: S29–S38.

MacKenzie, C. L. Jr. (2008) "History of the Bay Scallop, *Argopecten irradians*, Fisheries and Habitats in Eastern North America, Massachusetts through Northeastern Mexico." *Marine Fisheries Review* 70(3–4): 1–5.

Malthus, T. (1798) *An Essay on the Principle of Population*. London: J. Johnson, in St. Paul's Church-Yard. Available at Electronic Scholarly Publishing Project, http://www.esp.org.

Marks, A. R. (2011) "A Conversation with P. Roy Vagelos." *Annual Review Conversations. Annual Review of Biochemistry*. Available at http://www.annual reviews.org.

Martin, G. S. (1970) "Rous Sarcoma Virus: A Function Required for the Maintenance of the Transformed State." *Nature* 277(5262): 1021–1023.

——— (2004) "The Road to Src." *Oncogene* 23: 7910–7917.

Marshall, K. N, N. T. Hobbs, and D. J. Cooper (2013) "Stream Hydrology Limits Recovery of Riparian Ecosystems after Wolf Reintroduction." *Proceedings of the Royal Society, Series B* 280(1756): 20122977.

McLaren, B. E., and R. O. Peterson (1994) "Wolves, Moose, and Tree Rings on Isle Royale." *Science* 266: 1555–1558.

McNaughton, S. J. (1979) "Grazing as an Optimization Process: Grass-Ungulate Relationships in the Serengeti." *American Naturalist* 113(5): 691–703.

Mduma, S.A.R., A.R.E. Sinclair, and R. Hilborn (1999) "Food Regulates the Serengeti Wildebeest: A 40-Year Record." *Journal of Animal Ecology* 68: 1101–1122.

Mead, F. S. (1921) *Harvard's Military Record in the World War*. Boston: Harvard Alumni Association.

Michalak, A. M., E. J. Anderson, D. Beletsky, S. Boland, et al. (2013) "Record-Setting Algal Bloom in Lake Erie Caused by Agricultural and Meteorological Trends Consistent with Expected Future Conditions." *Proceedings of the National Academy of Sciences USA* 110(16): 6448–6452.

Monod, J. (1942) "Recherches sur la Croissance des Populations Bactèriennes." Paris: Hermann & Cie.

Monod, J., and F. Jacob (1961) "General Conclusions: Teleonomic Mechanisms in Cellular Metabolism, Growth, and Differentiation." *Cold Spring Harbor Symposia on Quantitative Biology* 26: 389–401.

Monod, J., G. Cohen-Bazire, and M. Cohn (1951) "Sur la Biosynthese de la β-Galactosidase (Lactase) chez *Escherichia coli*. La Specificite de l'Induction." *Biochimica et Biophysica Acta* 7: 585–599.

Morley, R., and I. Convery (2014) "Restoring Gorongosa: Some Personal Reflections." In I. Convery, G. Corsane, and P. Davis (eds.), *Displaced Heritage: Responses to Disaster, Trauma, and Loss*. Woodbridge, England: Boydell Press: 129–141.

Myers, R. A., J. K. Baum, T. D. Shepherd, S. P. Powers, and C. H. Peterson (2007) "Cascading Effects of the Loss of Apex Predatory Sharks from a Coastal Ocean." *Science* 315: 1846–1850.

Nair, P. "Brown and Goldstein: The Cholesterol Chronicles." *Proceedings of the National Academy of Sciences USA* 110(37): 14829–14832.

National Institutes of Health (2012) "Morbidity and Mortality: 2012 Chart Book on Cardiovascular, Lung, and Blood Diseases." Bethesda, Maryland: National Institutes of Health: National Heart, Lung, and Blood Institute.

Neill, U. S., and H. A. Rockman (2012) "A Conversation with Robert Lefkowitz, Joseph Goldstein, and Michael Brown." *Journal of Clinical Investigation* 122(5): 1586–1587.

Nicholls, K. H. (1999) "Evidence for a Trophic Cascade Effect on North-Shore Western Lake Erie Phytoplankton Prior to the Zebra Mussel Invasion." *International Association for Great Lakes Research* 25(4): 942–949.

Norton-Griffiths, M. (1979) "The Influence of Grazing, Browsing, and Fire on the Vegetation Dynamics of the Serengeti." In A.R.E. Sinclair and M. Norton-Griffiths (eds.), *Serengeti, Dynamics of an Ecosystem*. Chicago: University of Chicago Press: 310–352.

Novick, A., and L. Szilard (1954) "Experiments with the Chemostat on the Rates of Amino Acid Synthesis in Bacteria." In E. J. Boell (ed.), *Dynamics of Growth Processes*. Princeton, New Jersey: Princeton University Press: 21–32.

Nowell, P. C. (2007) "Discovery of the Philadelphia Chromosome: A Personal Perspective." *Journal of Clinical Investigation* 117(8): 2033–2035.

Nowell, P. C., and D. A. Hungerford (1960) "A Minute Chromosome in Human Chronic Granulocytic Leukemia." In "National Academy of Sciences. Abstracts of Papers Presented at the Autumn Meeting, 14–16 November 1960, Philadelphia, Pennsylvania." *Science* 132: 1497.

Obenour, D., D. Gronewald, C. Stow, and D. Scavia (2014) "2014 Lake Erie Harmful Algal Bloom (HAB) Experimental Forecast: This Product Represents the First Year of an Experimental Forecast Relating Bloom Size to Total Phosphorus Load." http://www.glerl.noaa.gov/res/Centers/HABS/lake_erie_hab/LakeErieBloomForecastRelease071514.pdf.

Paarlberg, D., and P. Paarlberg (2000) *The Agricultural Revolution of the 20th Century*. Ames: Iowa State University Press.

Paine, R. T. (1963a) "Ecology of the Brachiopod *Glottidia pyramidata*." *Ecological Monographs* 33(3): 187–213.

———— (1963b) "Trophic Relationships of 8 Sympatric Predatory Gastropods." *Ecology* 44(1): 63–73.

———— (1966) "Food Web Complexity and Species Diversity." *American Naturalist* 100(910): 65–75.

———— (1971) "A Short-Term Experimental Investigation of Resource Partitioning in a New Zealand Rocky Intertidal Habitat." *Ecology* 52(6): 1096–1106.

———— (1974) "Intertidal Community Structure." *Oecologia* 15: 93–120.

———— (1980) "Food Webs: Linkage, Interaction Strength and Community Infrastructure." *Journal of Animal Ecology* 49(3): 666–685.

———— (1992) "Food-Web Analysis through Field Measurement of per capita Interaction Strength." *Nature* 355: 73–75.

———— (2010) "Food Chain Dynamics and Trophic Cascades in Intertidal Habitats." In J. Terborgh and J. A. Estes (eds.), *Trophic Cascades: Predators, Prey, and the Changing Dynamics of Nature*. Washington, DC: Island Press: 21–36.

———— (2011) "Inspiration." In M. H. Graham, J. Parker, and P. K. Dayton (eds.), *The Essential Naturalist: Timeless Readings in Natural History*. Chicago: University of Chicago Press: 7–15.

Paine, R. T., and R. L. Vadas (1969) "The Effects of Grazing by Sea Urchins, *Strongylocentrotus* spp., on Benthic Algal Populations." *Limnology and Oceanography* 14(5): 710–719.

Pardee, A. B. (2003) "The Pajama Experiment." In A. Ullmann (ed.), *Origins of Molecular Biology: a Tribute to Jacques Monod*. Revised ed. Washington, DC: ASM Press.

Pearce, T. (2010) "'A Great Complication of Circumstances'—Darwin and the Economy of Nature." *Journal of the History of Biology* 43: 493–528.

Peterson, C. H., F. J. Fodrie, H. C. Summerson, and S. P. Powers (2001) "Site-Specific and Density-Dependent Extinction of Prey by Schooling Rays: Generation of a Population Sink in Top-Quality Habitat for Bay Scallops." *Oecologia* 129(3): 349–356.

Power, M. E., W. J. Matthews, and A. J. Stewart (1985) "Grazing Minnows, Piscivorous Bass, and Stream Algae: Dynamics of a Strong Interaction." *Ecology* 66(5): 1448–1456.

Rea, P. A. (2008) "Statins: From Fungus to Pharma." *American Scientist* 96(5): 408.

Reid, R. S. (2012) *Savannas of Our Birth*. Berkeley: University of California Press.

Riggio, J., A. Jacobson, L. Dollar, H. Bauer, et al. (2013) "The Size of Savannah Africa: A Lion's (*Panthera leo*) View." *Biodiversity and Conservation* 22: 17–35.

Rinta-Kanto, J. M., A.J.A. Ouellette, G. L. Boyer, M. R. Twiss, et al. (2006) "Quantification of Toxic *Microcystis* spp. during the 2003 and 2004 Blooms in Western Lake Erie Using Quantitative Real-Time PCR." *Environmental Science & Technology* 39(11): 4198–4205.

Ripple, W. J., and R. L. Beschta (2005) "Linking Wolves and Plants: Aldo Leopold on Trophic Cascades." *BioScience* 55(7): 613–621.

——— (2007) "Restoring Yellowstone's Aspen with Wolves." *Biological Conservation* 138: 514–519.

——— (2012) "Trophic Cascades in Yellowstone: The First 15 Years after Wolf Reintroduction." *Biological Conservation* 145(1): 205–213.

Ripple, W. J., and E. J. Larsen (2000) "Historic Aspen Recruitment, Elk, and Wolves in Northern Yellowstone National Park, USA." *Biological Conservation* 95: 361–370.

Ripple, W. J., T. P. Rooney, and R. L. Beschta (2010) "Large Predators, Deer, and Trophic Cascades in Boreal and Temperate Ecosystems." In J. Terborgh and J. A. Estes (eds.), *Trophic Cascades: Predators, Prey, and the Changing Dynamics of Nature*. Washington, DC: Island Press: 141–161.

Roeder, P. L. (2011) "Rinderpest: The End of Cattle Plague." *Preventative Veterinary Medicine* 102: 98–106.

Roman, J., M. M. Dunphy-Daly, D. W. Johnston, and A. J. Read (2015) "Lifting Baselines to Address the Consequences of Conservation Success." *Trends in Ecology & Evolution* 30(6): 299–302.

Rous, P. (1910) "A Transmissible Avian Neoplasm. (Sarcoma of the Common Fowl.)" *Journal of Experimental Medicine* 12(5): 696–705.

Rowley, J. D. (1973) "A New Consistent Chromosomal Abnormality in Chronic Myelogenous Leukaemia Identified by Quinacrine Fluorescence and Giemsa Staining." *Nature* 243: 290–293.

Rowley, J. D., H. M. Golomb, and C. Dougherty (1977) "15/17 Translocation, a Consistent Chromosomal Change in Acute Promyelocytic Leukaemia." *Lancet* 309(8010): 549–550.

Scandanavian Simvastin Survival Study Group (1994) "Randomised Trial of Cholesterol Lowering in 4444 Patients with Coronary Heart Disease: The Scandinavian Simvastatin Survival Study (4S)." *Lancet* 344(8934): 1383–1389.

Schoenly, K. G., J. E. Cohen, K. L. Heong, G. S. Arida, et al. (1996) "Quantifying the Impact of Insecticides on Food Web Structure of Rice-Arthropod Populations in a Philippine Farmer's Irrigated Field: A Case Study." In G. A. Polis and K. O. Winemiller (eds.), *Food Webs: Integration of Patterns and Dynamics*. New York: Chapman & Hall: 343–351.

Seuss, Dr. (1971) *The Lorax*. New York: Random House.

Sinclair, A.R.E. (1973a) "Population Increases of Buffalo and Wildebeest in the Serengeti." *African Journal of Ecology* 11(1): 93–107.

——— (1973b) "Regulation, and Population Models for a Tropical Ruminant." *African Journal of Ecology* 11: 307–316.

——— (1974) "The Natural Regulation of Buffalo Populations in East Africa." *African Journal of Ecology* 12: 185–200.

——— (1977) *The African Buffalo: A Study of Resource Limitation of Populations.* Chicago and London: University of Chicago Press.

——— (1979) "The Eruption of the Ruminants." In A.R.E. Sinclair and M. Norton-Griffiths, *Serengeti, Dynamics of an Ecosystem.* Chicago: University of Chicago Press: 82–103.

——— (2003) "Mammal Population Regulation, Keystone Processes and Ecosystem Dynamics." *Philosophical Transactions of the Royal Society Series B* 358: 1729–1740.

——— (2012) *Serengeti Story.* Oxford: Oxford University Press.

Sinclair, A.R.E., and C. J. Krebs (2002) "Complex Numerical Responses to Top-Down and Bottom-Up Processes in Vertebrate Populations." *Philosophical Transactions of the Royal Society Series B* 357: 1221–1231.

Sinclair, A.R.E., and K. L. Metzger (2009) "Advances in Wildlife Ecology and the Influence of Graeme Caughley." *Wildlife Research* 36: 8–15.

Sinclair, A.R.E., and M. Norton-Griffiths (1979) *Serengeti, Dynamics of an Ecosystem.* Chicago: University of Chicago Press.

Sinclair, A.R.E, S. Mduma, and J. S. Brashares (2003) "Patterns of Predation in a Diverse Predator-Prey System." *Nature* 425: 288–290.

Sinclair, A.R.E., K. L. Metzger, J. S. Brashares, A. Nkwabi, et al. (2010) "Trophic Cascades in African Savanna: Serengeti as a Case Study." In J. Terborgh and J. A. Estes (eds.), *Trophic Cascades: Predators, Prey, and the Changing Dynamics of Nature.* Washington, DC: Island Press: 255–274.

Slobodkin, L. B. (2009) "My Complete Works and More." *Evolutionary Ecology Research* 11: 327–354.

Smith, D. W. (2005) "Ten Years of Yellowstone Wolves, 1995–2005." *Yellowstone Science* 13: 7–33.

Sogawa, K. (2015) "Planthopper Outbreaks in Different Paddy Ecosystems in Asia: Man-Made Hopper Plagues that Threatened the Green Revolution in Rice." In K. L. Heong, J. Cheng, and M. M. Escalada (eds.), *Rice Planthoppers: Ecology, Management, Socio Economics and Policy.* Dordrecht: Springer: 33–63.

Southwood, R., and J. R. Clarke (1999) "Charles Sutherland Elton, 29 March 1900–1 May 1991." *Biographical Memoirs of Fellows of the Royal Society* 45: 130–146.

Spector, D. H., H. E. Varmus, and J. M. Bishop (1978) "Nucleotide Sequences Related to the Transforming Gene of Avian Sarcoma Virus Are Present in DNA of Uninfected Vertebrates." *Proceedings of the National Academy of Sciences USA* 75(9): 4102–4106.

Spinage, C. A. (2003) *Cattle Plague: A History.* New York: Kluwer Academic/Plenum.

Stalmans, M., M. Peel, and T. Massad (2014) "Aerial Wildlife Count of the Parque Nacional da Gorongosa, Mozambique, October 2014." Report for Gorongosa National Park.

Starling, E. H. (1923) "The Wisdom of the Body: The Harveian Oration, Delivered before The Royal College of Physicians of London on St. Luke's Day, 1923." *British Medical Journal* 2(3277): 685–690.

Stehelin, D., H. E. Varmus, and J. M. Bishop (1976) "DNA Related to the Transforming Gene(s) of Avian Sarcoma Viruses Is Present in Normal Avian DNA." *Nature* 260: 170–173.

Stent, G. S. (1985) "Thinking in One Dimension: The Impact of Molecular Biology on Development." *Cell* 40: 1–2.

Stolzenburg, W. (2009) *Where the Wild Things Were: Life, Death, and Ecological Wreckage in a Land of Vanishing Predators.* New York: Bloomsbury USA.

Summerhayes, V. S., and C. S. Elton (1923) "Contributions to the Ecology of Spitsbergen and Bear Island." *Journal of Ecology* 11(2): 214–286.

Talbot, L. M., and D.R.M. Stewart (1964) "First Wildlife Census of the Entire Serengeti-Mara Region, East Africa." *Journal of Wildlife Management* 28(4): 815–827.

Thornburn, C. (2015) "The Rise and Demise of Integrated Pest Management in Rice in Indonesia." *Insects* 6: 381–408.

Tobert, J. A. (2003) "Lovastatin and Beyond: The History of the HMG-COA Reductase Inhibitors." *Nature Reviews* 2: 517–526.

Tracy, S. W. (2012) "The Physiology of Extremes: Ancel Keys and the International High Altitude Expedition of 1935." *Bulletin of the History of Medicine* 86(4): 627–660.

Tucker, J. B. (2001) *Scourge: The Once and Future Threat of Smallpox.* New York: Atlantic Monthly Press.

Ullmann, A. (2003) *Origins of Molecular Biology: A Tribute to Jacques Monod.* Revised ed. Washington, DC: ASM Press.

Umbarger, H. E. (1956) "Evidence for a Negative-Feedback Mechanism in the Biosynthesis of Isoleucine." *Science* 123: 848.

——— (1961) "Feedback Control by Endproduct Inhibition." *Cold Springs Harbor Symposia on Quantitative Biology* 26: 301–312.

US Fish and Wildlife Service (1994) "Final Environmental Impact Statement: The Reintroduction of Gray Wolves to Yellowstone National Park and Central Idaho." Helena, Montana: US Department of the Interior.

——— (2000) "Federal Aid in Sport Fish Restoration Handbook." Fourth ed. Washington, DC: US Department of the Interior.

Van der Kloot, W. (2010) "William Maddock Bayliss's Therapy for Wound Shock." *Notes and Records: The Royal Society Journal of the History of Science* 64: 271–286.

Vanni, M. J., C. Luecke, J. F. Kitchell, Y. Allen, et al. (1990) "Effects on Lower Trophic Levels of Massive Fish Mortality." *Nature* 344: 333–335.

Varmus, H. (2009) *The Art and Politics of Science*. New York: W. W. Norton & Company.

Vogelstein, B., N. Papadopoulos, V. E. Velculescu, S. Zhou, et al. (2013) "Cancer Genome Landscapes." *Science* 339: 1546–1558.

Wapner, J. (2013) *The Philadelphia Chromosome: A Mutant Gene and the Quest to Cure Cancer at the Genetic Level*. New York: The Experiment.

Wasserman, E. (2000) *The Door in the Dream: Conversations with Eminent Women in Science*. Washington, DC: Joseph Henry Press.

Wells, H. G. (1927) *Meanwhile: The Picture of a Lady*. London: E. Benn.

Wentzel, V. (1964) "Mozambique: Land of the Good People." *National Geographic* 126(2): 197–231.

Wheeler, S. (2006) *Too Close to the Sun: The Audacious Life and Times of Denys Finch Hatton*. New York: Random House.

White, P. J., D. W. Smith, J. W. Duffield, M. Jimenez, et al. (2005) "Wolf EIS Predictions and Ten-Year Appraisals." *Yellowstone Science* 13(1): 34–41.

White, S. E. (1915) *The Rediscovered Country*. Garden City, New York: Doubleday, Page & Company.

Williams, R. (2010) "Joseph Goldstein and Michael Brown: Demoting Egos, Promoting Success." *Circulation Research* 106(6): 1006–1010.

Wolfe, E. L., A. C. Barger, and S. Benison (2000) *Walter B. Cannon, Science and Society*. Cambridge, Massachussetts: Boston Medical Library.

Worster, D. (1994) *Nature's Economy: A History of Ecological Ideas*. Second ed. Cambridge: Cambridge University Press.

Yim, E., and J. Park (2005) "The Role of HPV E6 and E7 Oncoproteins in HPV-Associated Cervical Carcinogenesis." *Cancer Research and Treatment* 37(6): 319–324.

Yong, E. (2013) "Dynasty." *Nature* 493: 286–289.

Zech, L., U. Haglund, K. Nilsson, and G. Klein (1976) "Characteristic Chromosomal Abnormalities in Biopsies and Lymphoid-Cell Lines from Patients with Burkitt and Non-Burkitt Lymphomas." *International Journal of Cancer* 17(1): 47–56.

INDEX

Page numbers followed by "f" indicate figures and images.

A CONVERSATION WITH SEAN B. CARROLL, AUTHOR OF *THE SERENGETI RULES: THE QUEST TO DISCOVER HOW LIFE WORKS AND WHY IT MATTERS*

1. Why did you write this book?

The quest to understand how nature works has been one of the great challenges—and triumphs—of modern times. I wanted to share the often little-known stories of some of the pioneers who have penetrated some of the deepest mysteries of life—from how the human body works to how whole communities of creatures interact. This knowledge empowers us to improve the quantity and quality of life across the planet.

2. Your book is called *The Serengeti Rules*. What are those rules?

Just as there are rules that regulate the numbers of different kinds of molecules and cells in the body, there are ecological rules that regulate the numbers and kinds of animals and plants in a given place. I have called these the "Serengeti Rules" because that is one place where they have been worked out and they determine, for example, how many lions, or buffalo, or elephants live on an African savanna. But these rules apply all over the globe, in oceans, rivers, and lakes, as well as on land.

3. The scientists portrayed in *The Serengeti Rules* are admirable, sometimes heroic figures. Why did you choose to organize the book around their stories?

I am a firm believer in the power of stories. We learn better from stories because they help us connect series of events and to understand cause and effect in the world. And we are more engaged when human characters are central to a story. Science is far more enjoyable, understandable, and memorable when we follow scientists all over the world and share in their struggles and triumphs.

4. In many ways, your book bridges the divide—and helps unify—two fields of biology: molecular biology and ecology. Why is this important?

Biology is central to the future of humanity, but molecular biology and ecology have long been separated—both in the curriculum and in how biology is organized on campuses and by funding agencies. The sad result is that the two communities do not know much about each others' priorities and accomplishments. That is a shame and an obstacle at this moment in time, when the fate of the biosphere is in jeopardy. I have tried to present the simple case that ecology is to planetary health what molecular biology is to human health—the critical knowledge for building a better future.

5. You use an analogy from sports to explain how scientists have figured out how to treat many diseases. How does that analogy apply to medicine?

In the body, the key "players" are molecules that regulate a process. To intervene in a disease, we need to know what players are injured or missing or what rules of regulation have been broken. The task for biologists is to identify the important players in a process, figure out the rules that regulate their action, and then design medicines that target the key players. In the book, I tell the stories of just how that was done to make such dramatic progress against heart disease and cancer.

6. Does the same analogy apply to the health of the planet—to conserving and restoring ecosystems?

Absolutely. We need to know the key species in any given community and the rules that govern their interactions with other species. But in contrast to the considerable care and expense we gladly undertake in applying molecular rules to human medicine, we have done a very poor job in considering and applying ecological rules to human affairs.

7. But as you describe in several chapters, there have been some encouraging successes in restoring species and habitats.

Yes, and I thought it was very important to tell those stories, to show that even war-torn and devastated places like Gorongosa National Park in Mozambique could rebound given time, protection, and the efforts of just a small band of extraordinarily dedicated people.

8. You visited Gorongosa in the course of writing this book. What was that experience like?

Life-changing. The people behind the Gorongosa Restoration Project are so inspiring, and the magnitude of the recovery in just ten years is astounding and so encouraging. If Gorongosa can be rescued from utter disaster, we should all take heart that we can restore other places and species. We are a clever species—we have made huge strides against human diseases in a short time span. We need the same sort of drive and sense of urgency applied to the health of the planet we depend upon.

9. For readers who want to use your book as a teaching tool in a classroom setting, what do you suggest?

The overarching theme of the book is that everything in the living world is regulated. Thus regulation, like evolution, should be a cross-cutting and unifying theme in biology. This theme and these stories may help students see relationships between different scales in biology. Every cell contains a society of molecules, every organ a society of cells, every body a society of organs, every habitat a society of organisms. Understanding the interactions within each of those societies are the primary aims of molecular biology, physiology, and ecology.

10. What do you hope students in particular will take away from *The Serengeti Rules*?

First, I hope that they feel inspired by the stories of some exceptional people who tackled and solved great mysteries. Second, that they feel enriched with fresh insights into the wonders of life at different scales. Third, that they feel more hope for the future—that there is time to change the road we're on.

For additional resources for using
The Serengeti Rules in the classroom please see

www.press.princeton.edu/releases/
serengeti-rules-course-material.pdf.